This Working-Day World

*Due Fri
after
Spring
break

Feminist Perspectives on The Past and Present Advisory Editorial Board

This Working-Day World
Women's Lives and Culture(s) in Britain 1914–1945

Edited by

Sybil Oldfield

Taylor & Francis
Publishers since 1798

UK Taylor & Francis Ltd, 4 John St., London WC1N 2ET
USA Taylor & Francis Inc., 1900 Frost Road, Suite 101, Bristol, PA 19007

First published 1994

A Catalogue Record for this book is available from the British Library

ISBN 0 7484 0107 5
ISBN 0 7484 0108 3 pbk

Library of Congress Cataloging-in-Publication Data are available on request

Series cover design by Amanda Barragry, additional artwork by Hybert • Design & Type. Photograph, *Stall-holder at Brixton Market*, by Margaret Monck.

Typeset in 10/11pt Times
by Graphicraft Typesetters Ltd., Hong Kong

Printed in Great Britain by Burgess Science Press, Basingstoke on paper which has a specified pH value on final paper manufacture of not less than 7.5 and is therefore 'acid free'.

Contents

Contents

Illustrations

Acknowledgments

For permission to reproduce the illustrations we are most grateful to the following:

To John Monck for 'Stall-holder at Brixton Market' (cover), photograph by Margaret Monck.

To Tony Cripps for the photograph of Marian Ellis.

To The Wiener Library, London for 'Jewish refugee children at UK customs'.

To Jenny Granger-Taylor and the British Library for *Allegory* (1934) by Edith Granger-Taylor.

To the Imperial War Museum for *Ruby Loftus Screwing a Breech-Ring* by Dame Laura Knight.

To Marian Kratochwil and the Tate Gallery for *Vanessa* (1937) by Dame Ethel Walker.

To the Lee Miller Archives, Chiddingly, East Sussex, for *Revenge on Culture* by Lee Miller.

Mrs Phyllis Smith at the University of Sussex was indispensable in the production of a publishable text – we all owe her our sincerest thanks.

Rosalind: O how full of briers is this working-day world!

SHAKESPEARE, *As You Like It*

Introduction

Sybil Oldfield

Every generation feels it is living in terrible times:

'We have a Government bent on destroying the foundations on which, not so much the greatness, but the very existence of our State depends. . . .'
'I agree with you from the bottom of my heart,' said Mr. Sykes. 'I see no gleam of light anywhere.' Then the two old friends felt happier, and went indoors. (F.M. Mayor, *The Rector's Daughter*, 1924)

Yet surely those who were adults between 1914 and 1945 had more reason than any generation this century for despair. They had to endure two world-wide competitions in massacre separated by a world-wide economic slump and the triumph of dictatorships. Nevertheless, this is not a gloomy book. British women's history during this period is not one of passive suffering – far from it. The following essays demonstrate women's insistence on survival and resistance; they focus on women's creativity in the construction of counter-cultures; and they give due credit to the taken-for-granted, life-enabling *work* of women in the background, whether that work be the weekly wash, or finding foster homes for refugee children or doing obscure but vital medical research. Our emblematic cover photograph shows two women – perhaps the two faces of women in Britain 1914–1945 – one anxious and worn, the other laughing and ready to take on the world.

I think this book is unusual in two ways. First, it looks at more aspects of life than those usually included under 'history', since we include house-work, the visual arts, broadcasting, literature, and science. Secondly, it refuses to confine history to the interaction of individuals in power or to the inter-action of powerful groups or to the interaction between individuals and powerful groups. Here are several individuals not in power – isolated pacifists in World War One, a single mother in World War Two, a teenage refugee in Manchester – for history is lived singly as well as collectively and it is *felt* singly more often than it is felt collectively. Whether or not such individual accounts 'deserve a place in the story' is up to the reader to judge. Do we deserve a place in history only if we are clearly representative of others – and of how many others do we have to be representative?

Do British women 1914–1945 need or deserve a history separate from men at all? Yes, insofar as women had and have gender-specific experience. No man could have been in the predicament of Doreen Bates. And yes, insofar as there were important women-only groups at this time such as the Women's Co-operative Guild, the Women's Institutes, the Women's Teacher Training Colleges and the feminist Six Point Group. But there is a way in which women's history has been artificially and unnecessarily separated from that of men in that women have so often simply been left out. A Martian could read C.L. Mowat's *Britain Between the Wars 1918–1940* (1955) or Julian Symons' *The Thirties: A Dream Revolved* (1960) or Pimlott's *Labour and the Left in the 1930s* (1978) or Stevenson and Cook's *The Slump* (1978) among many other such works and conclude that the population of Britain between 1914 and 1945 must have been 90 per cent male. Indeed, given that the index to Mowat's 698 pages lists the names of 600 men and only 25 women that percentage is nearer 95 per cent male. And that is socio-political history, not military or naval or colonial history which are 100 per cent male. The explanation for such an extraordinary under-representation of women lies not just in the sex of the historian but also in the historian's conception of history. If history, by definition, be held to be the history of individuals or groups with power, then women remain unnoticed.[1] But what, then, do we mean by power? A book such as this, focusing principally on many aspects of women's working lives, exists not only to redress a gross imbalance in the imaging of our shared past, but also to ask what in our past we hold to be worth remembering. Separate women's history is a necessary stage not just in the reconceptualizing of history but of society and life itself.

This Working-Day World is feminist in that it insists that women also live and work in the world and that our experience matters just as much as does that of men. Our book is an empirical, humanist–feminist text.[2] Women are seen to be as human as men, but not better than men – nor simply oppressed by men in the home or outside. Most of us, like most men, are full of limitations, blind spots, debilitating insecurity, and a compensatory need to control someone or something. The following essays are not, therefore, a mere series of panegyrics; the seduction of some British women by the militaristic nationalism, authoritarianism, and racism of the period is fully documented. And, as is the case with all serious history, the more one knows, the less easy it becomes to generalize.

We realize that we have not produced a comprehensive overview of the period, merely tried to tell some untold stories.[3] Where are the Welsh women, Irish women in Britain, women of colour, women journalists, women musicians, women MPs, women film-makers, women pensioners, women trades unionists, nurses, charwomen, actresses, or shorthand-typists of 1914–1945? These histories are still waiting to be told, and probably need to be addressed by a careful collation of regional and urban case studies, and, above all, by oral history – before it is too late.

Notes

1   Beatrix Campbell located a classic example of women's invisibility to male chroniclers of the period in Orwell's *The Road to Wigan Pier*; see her *Wigan Pier Revisited* (1984), London, Virago, p. 101.

2   Cf. Deborah Thom, 'A Lop-Sided View: Feminist History or the History of Women', p. 47 in K. Campbell (1992), *Critical Feminism*, Buckingham, Open University Press: empirical studies reject 'assumptions that women are a unitary category, that women are victims, that womanhood as a concept determines – rather than that their own activity shapes – their lives'.

3   Other attempts to tell hitherto untold stories, not listed elsewhere in this book, are listed below.

Further Reading

ADAM, R. (1975) *A Woman's Place, 1910–1975*, London, Chatto.

ALBERTI, J. (1989) *Beyond Suffrage*, London, Macmillan.

ALEXANDER, S. (Ed.) (1988) *Women's Fabian Tracts*, London, Routledge.

BEAUMAN, N. (1983) *A Very Great Profession: The Woman's Novel 1914–1939*, London, Virago.

BINDMAN, L., BRADING, A. and TANSEY, T. (Eds) (1993) *Women Physiologists*, Colchester Portland Press.

BLACK, N. (1989) *Social Feminism*, Ithaca, Cornell University Press.

BRAYBON, G. and SUMMERFIELD, P. (1987) *Out of the Cage*, London, Pandora.

BRIGGS, S. (1975) *Keep Smiling Through*, London, Weidenfeld.

CALDECOTT, L. (1984) *Women of our Century* (including Naomi Mitchison, Dora Russell, Janet Vaughan and Barbara Wootton), London, Ariel.

CALDER, A. and SHERIDAN, D. (Eds) (1985) *Speak for Yourself: A Mass-Observation Anthology 1937–1949*, Oxford, Oxford University Press.

DAVIES, A. (1992) *Leisure, Gender and Poverty: Working-Class Culture in Salford and Manchester*, Buckingham, Open University Press.

DYHOUSE, C. (1989) *Feminism and the Family*, Oxford, Basil Blackwell.

GLUCKSMAN, M. (1990) *Women Assemble: Women Workers and the New Industries*, London, Macmillan.

HALL, R. (1977) *Marie Stopes*, London, Deutsch.

INGRAM, A. and PATAI, D. (Ed.) (1993) *Rediscovering Forgotten Radicals – British Women Writers 1889–1939*, Chapel Hill and London, University of North Carolina Press.

JAFFE, P. (1988) *Women Engravers*, London, Virago.

LIGHT, A. (1991) *Forever England: Femininity, Literature and Conservatism between the Wars*, London, Routledge.

MELMAN, B. (1988) *Women and the Popular Imagination in the Twenties*, London, Macmillan.

PENROSE, A. (1993) *Lee Miller's War*, London, Thames and Hudson.

RILEY, D. (1983) *War in the Nursery: Theories of the Child and Mother*, London, Virago.

SMITH, H. (Ed.) (1990) *British Feminism in the Twentieth Century*, London, Edward Elgar.

SUMMERFIELD, P. (1984) *Women Workers in the Second World War*, London, Croom Helm.

TYLEE, C. (1990) *The Great War and Women's Consciousness*, London, Macmillan.

WILLIAMS, V. (1986) *The Other Observers: Women Photographers in Britain*, London, Virago.

Section I

Social History

Chapter 1

The Weekly Wash

Christine Zmroczek

There are many people still living for whom the period from 1914 to 1945 is part of their own remembered past, and many many more for whom at least a part of that period is within their own lives. Still more of us have this period as a part of our family memories; family and friends talk about their past, their childhoods and younger lives, and especially of their wartime experiences. There can never, of course, be a complete picture of a historical period, not even a period of which people can speak from their own memories. Those familiar with the methodologies of oral history, feminist sociology and ethnography,[1] will be well aware of the debates on the distortions and selectivities of memory, the re-presentation of the self as a 'moveable feast'. The self is not fixed but re-constructed over time and the particular construction at any one time depends on the situation, the person who is listening, the questions asked, and the relationships between the speaker/rememberer and the listener/s. That being said it has been, for me, particularly exciting to ask women questions about the minutiae of life which so often do not appear in traditional history, nor in the documents available for its study.

I have carried out in-depth interviews in Brighton with twenty women over 65 and talked with about fifty others in groups or in less formal ways or on specific issues. I asked them about their memories of how housework was done in their childhood, in the 1920s and 1930s, although a few of the older women described childhoods before that, and how they themselves did the work as adults, in the 1940s and 1950s. I wanted to examine how and when domestic technologies of various kinds made a difference in women's lives, and what kind of a difference.

I focused on washing – one of the most detested but most implacable of tasks. Washing methods and technologies in the early part of the inter-war period, I learned, have remained pretty much the same as fifty or even a hundred years before for most women of all classes in Britain.[2] The main difference between the classes, as always, was that women who could afford it paid someone else to do the work – a servant, washerwoman or commercial laundry worker. And it was working-class women who did these jobs, who did the washing for the other classes as well as for their own households – in many cases 'taking in' other people's washing into their homes as paid work.

I begin with a brief summary of some details of the childhood homes of three of the women I interviewed in order to illustrate how working-class women in Brighton lived and worked in the 1920s and 1930s. Their stories are typical.

Audrey, Elsie and Olive

Audrey and Elsie were born in 1926 and Olive in 1921. Audrey's mother had at least one and perhaps several miscarriages and Audrey had only one sister, nine years younger than she. When Audrey was about 3 her parents moved from the two top-floor rooms they rented in a tall house, to take half of the house rented by Audrey's great-grandparents, in which they shared the kitchen, scullery, lavatory outside, and small garden. Each had two rooms upstairs and one downstairs. There was gas light and a gas stove, a copper and a range in the kitchen. They used coal and wood on these, as Audrey's great-grandfather had worked for a coal merchant and had a supply of coal allowed him. They kept a mangle outside in the garden shed. The women cooked and kept house separately. Audrey's father was a window dresser and later a roundsman for a large tea company. Her mother had been a secretary before she had children and she returned to this work during World War Two.

Elsie had seven brothers and one sister. Some of the brothers slept at a neighbour's house down the street and Elsie and her sister slept next door in their grandmother's bed as their parents had only a kitchen/scullery and one other room in the small terraced house they shared with two other families. Elsie's grandmother shared her house with another large family, and the women of Elsie's family carried on a business as washerwomen using Grandma's kitchen/scullery to work in and to dry clothes. The houses had gas lighting but not in all the rooms. Cooking was done on the kitchen range, and there was a copper in Grandma's house, in which they did the washing. Elsie's childhood tasks included collecting rubbish to burn on the copper fire from local shopkeepers or from stallholders at the nearby market. The houses had no gardens but had a small yard at the back. This was sometimes used for drying the laundry, but mostly drying was done indoors to avoid smuts or soiling and where there was always a fire to speed up the process. Elsie's father was a road sweeper, and her mother besides the laundry work also did childminding for women in the street who had market stalls, and she laid out the dead, following in the tradition of her mother. She and Elsie's grandmother were also involved in illegal betting. Elsie recalled the money for the bets being passed through the grating in the street into the window of the basement kitchen where the women worked.

Olive was the youngest in her family and had one sister and two brothers. The whole family including mother and father lived in three basement rooms in a terraced house, with a scullery but with no garden or yard. Washing was dried in the kitchen 'with lines up'. They had gas lighting in some rooms, but cooking was done on the fire, with pots hung over it, as there was never enough fuel for the range. Like Elsie, Olive used to collect rubbish to burn in it, but she recalls that there was never enough to have a really big fire. Olive's father was an ex-army corporal and was out of work for most of

Olive's early years. In the late 1920s he obtained some steady but seasonal employment with the council parks and gardens. Her mother went out to clean, took in washing, did waitressing and odd tasks for a small local family-run laundry, all depending on her state of health which was very poor.

The Washing

The main essentials needed to do the washing are copious supplies of water and energy. The energy has usually (and until quite recently) been provided almost entirely by women. The water in the inter-war years, even in urban areas, was not necessarily readily available in every home. The classic study by Margery Spring Rice and the Women's Health Committee in the 1930s surveyed 1,250 working-class women from different backgrounds and from all parts of the country.[3] It revealed that many women had no private indoor water supply. Many had to walk some distance to obtain water from a communal tap, a pump, or a stand-pipe, and this often included going up and down several flights of stairs. This could also be true even if the water supply were indoors and not down the street – as was confirmed by the women I interviewed. Audrey, for example, said: 'They lived at the top of the house, a high storeyed house, and my mother told me when I was born it was very difficult because she had to carry all the water all the way up these flights of stairs as they had this top flat'.

We might just stop and think exactly what this must have meant for women in this situation when so many daily tasks require water – cooking and cleaning and washing babies and children as well as clothes. Every drop of water had to be carried both upstairs to be used and then down again when it was dirty and had to be disposed of. Heavy buckets and bowls were carried some distance many times a day, often with small children to supervise on the stairs as well. It was probably a cold, wet job too, if the water source were outside, and no doubt the carrier often spilt plenty of water on herself. Thus something as simple as obtaining and disposing of water involved very hard labour for many working-class British women at this time.

For some of my informants, moving to a new council house – 2,500 were built in Brighton in the inter-war years, and 1,250,000 nationally – provided their first taste of such luxuries as piped hot water and electricity. Olive was 10 when her family moved in 1931.

> We were thrilled with the house. Just imagine being able to have a bath in a special room, to be able to turn on the tap and get hot water. The most magical thing of all was to press a switch and the electric light came on. My mother hired an electric boiler, cooker, kettle, and iron from the Electricity Board. It was all brand new and marvellous for her. Best of all she could do the washing without lifting heavy buckets of water or lighting a fire under the boiler.

Relatively few working-class women were as fortunate as Olive's mother in respect of new electrical household appliances. In the 1930s nearly a third of homes still did not have electricity – and of those that did a majority were wired only for light and had no plugs for appliances.[4] The Heating of Dwellings Inquiry in 1942[5] surveyed over 5,000 working-class homes and found that less than a quarter of them had piped hot water. Another survey in the 1940s found that over a third of households nationally heated all their water for washing on the stove.[6]

What then were the main technologies used by women in the 1920s and 1930s when doing the washing? Aside from systems to provide water, electricity and gas, the copper and boiler and the mangle or wringer were amongst the most important items of washday technology in households in all classes and income levels during the inter-war years, and in fact until after the end of World War Two. The copper was for heating water and to boil up the wash (considered essential to get things clean). It was either a large built-in tub usually with a space underneath to light a fire to boil up the water or free standing and heated by gas or electricity. In either case there was the risk of scalding small children. The mangle was a heavy wood and iron structure with large wooden rollers turned by a handle to squeeze out the water in between rinses and before drying. It also smoothed the linens and this meant that less ironing was possible. It was often too big to fit indoors and was housed outside as in Audrey's home. A wringer was smaller and portable, usually with rubber rollers, and could be attached to a table or the sink and stored away after use. However, being less heavy it did not get as much water or as many creases out of the clothes. Together with a range of tubs or baths, buckets, basins and jugs, and usually a washboard, a scrubbing brush, and a cake of hard soap, these formed the basic equipment needed to do the work. For drying, a washing line – outdoors, indoors, or both – was needed, and some heat, the sun, a fire or gas. And finally there was the ironing, very important at a time when most fabrics were natural cotton or linen and much pride was invested in good ironing. Most women had two flat irons that they heated on the fire or gas and used alternately. A good deal of skill was attached to ironing in this way. It was also important to avoid soiling the linen at this final stage, and women used various cloths to wipe off any dirt from heating the iron. It had to be at the correct temperature and it cooled rapidly. Elsie's aunt was the ironing specialist in the family business and possessed a range of irons for different purposes.

> The clothing the servants wore, that was specialist work, even their pinafores, their little white pinafores they used to have fluted edges, and they all had to be – my aunt used to stand there with these, like a fork, a four-pronged fork which came down to a piece of metal shape, and she used to stand there, crimping away, slipping the fork in and out of the frills, and shoving these forks in the fire to heat them. She'd do all the shirts, too.

There were some new fabrics introduced in the 1920s and 1930s as a result of new technological processes, such as viscose and rayon. However, several of my informants recalled disasters when their clothing actually melted as a result of their traditional methods. Elsie hung her new viscose slip up to dry near the gas jet in the kitchen 'and it just disappeared'. Audrey told me:

I didn't have many nice things, but I did have a very very lovely silk slip that my mother had bought from the woman who sold second-hand clothes . . . and to my horror when I put the iron on my slip, I looked and I had burnt a great hole . . . my treasured possession . . . well of course that was my lesson which stuck in my mind forever, that when you're ironing you must test.

Even the minimal facilities for washing described above were not necessarily available in all working-class homes. Some women used public or communal facilities if they had access to them.[7] Brighton, for example, does not appear to have had a public wash-house, but many rows of terraces had shared wash-houses. Many women just had to make do somehow without. Lack of suitable equipment and cramped and crowded conditions made even more difficult the arduous work of washday at home as described to me by many women I interviewed.

Briefly, and with variations according to the facilities and equipment used, each batch of washing meant soaking, scrubbing, usually boiling, several rinses, several wringings, mangling, and hanging up to dry, followed later by ironing, airing, folding, and putting away. Each wash involved a great deal of physical exertion, much bending and lifting, and carrying of water and heavy wet washing. The washing had to be scrubbed, and then stirred and agitated by hand with a special stick or 'dolly' whilst in the copper. Then the heavy wet steaming clothes had to be hauled out of the boiling water with a long pole or tongs, a dangerous process. Blueing for whites and starching for some items were also considered essential, especially by the older generations. After this came rinsing and wringing either by hand or through a mangle or wringer, in either case hard work. Finally everything had to be carried outside to be hung out to dry or, if the weather were wet, hung up indoors wherever possible. After hanging out the clothes, the clearing away and cleaning up of the work area had to be done as the weekly wash had usually made everywhere wet, steamy, and soapy – especially the floor.

Many women said that their mothers, or they themselves in later years, later used the left-over water to wash all the floors and wash down the yard and street outside, because they did not like to waste anything, not even soapy water. Most women reported that the washing took at least a day, and often even longer. According to my informants it seems that some washing was done nearly every day in many households although the 'big' wash when the copper was heated took place once a week, usually on Mondays. There would also, of course, be drying and ironing on other days. Many said there were few occasions when there was not washing hanging around drying.

Were there no new technologies in the 1920s and 1930s which could alleviate the burdens of this arduous work? What about washing machines? Given the work and the equipment described above, surely women would have jumped at the chance to lighten the burden of washing mechanically? In 1938 just 4 per cent of UK households had a washing machine of any kind. Yet only three years later statistics show that 52 per cent of US families had access to one in their homes.[8]

Christine Zmroczek

A Short History of Washing Machines

The earliest mention of a machine for washing clothes appears in 1677 in the diary of Sir John Hoskyns, a Fellow of the Royal Society, and the first patent was taken out in 1691. In the eighteenth and nineteenth centuries designs for washing machines proliferated, although most were manufactured for industrial purposes and only a few were made for domestic use. Most of the early domestic machines were little more than wooden boxes or barrels which were filled manually and emptied with bowls or scoops. Some were on rockers to enable a gentle to and fro motion to be obtained by the women operating it by hand or foot. The foot-operated machines could be said to be amongst the first to be time-saving since they left the hands free to do other tasks such as mending, preparing vegetables, or rocking the baby. This rocking motion was incorporated into the popular wooden Faithful washer of 1906 and a later open galvanized tub model which was still sold in the 1940s.

Some earlier machines had a wooden board which was pressed up and down on the clothes or wooden blades or else a dolly on a horizontal pole inside the tub was rotated by a handle, and later by a lever, on the outside. This action imitated the traditional hand-held dolly which a woman simultaneously twisted and lifted up and down in a tub of washing to enable the water to pass through the clothes vigorously – at the cost of a great deal of the woman's energy. Women I interviewed describe their mothers using a dolly in their childhood, and some continued to use this method themselves until well after World War Two. Other machines, such as the highly popular series of Victress Vowel washing machines, used the same principle as the butter churn. They were still highly recommended by laundry manuals in the 1920s.

Electric motors were first attached to washing machines in the early 1900s. The electric power was used to move the dolly, thus replacing the woman's effort on the handle or lever, and also to operate the wringer attached to some models. By 1916 the electric motor was sealed in the bottom of the tub, which was much safer and more satisfactory, but still many improvements were needed to ensure really safe and efficient working.

New designs and technical innovations continued to appear throughout the inter-war years, although most electrically powered machines were imported from the US, Canada, or Germany and were very costly. Two surveys in the trade journal *The Electrical Review* in 1937 and 1939 illustrated both the types of electric washing machines available and the influence of North American companies who set up sales and assembly plants in Britain. Most machines were little more than an electrically heated wash boiler, perhaps with an electrically operated detachable wringer. Many of the designs described in the articles were still the round, galvanized tub type, usually by this time incorporating a lid. However, there is a photograph and lengthy description of the Bendix automatic machine claiming that all its functions were automatic so that it could be set and left to go through them. The 1939 article also described three different manufacturers' twin-tub machines which were to become so popular after World War Two.

So why did so few British women of any class have washing machines? The reasons are complex. The most obvious was the price. Washing machines, whether imported or home produced, remained expensive throughout the inter-war period. In the 1920s electric washers cost between £45 and £60 when the average salary for managers and professional men was between £500 and £700 a year. (Few women were in the professions.) Prices seem to have come down a little by the mid 1930s. The Hotpoint BTH electric washer sold in 1935 for £25 or £35 with a roller-iron fitted under it, and there were some more modestly priced ones at around £20. But those who could afford these prices could also pay others to do the work and the attitude may well have been 'why spend all that money to make things easier for a servant?' Non-electric, manually agitated machines were less expensive, but still beyond the reach of most people. The basic model of the Ewbank Master Clothes Washer, for example, cost £4 7s 6d at a time when a working-class man in full-time work was fortunate if he earned as much as £2 a week[9] and women, of course, were paid far less. A washerwoman, for example, worked a full day for 4s or 5s (25p) according to my informants, whilst one said that in her rural home 1s 6d (6p) was the rate.

For those aspiring to middle-class life but unable to afford a live-in servant, a washing machine may have seemed more attractive – and indeed many domestic appliances were advertised as servant substitutes.[10] However, most washing machines were still quite basic even if electrically powered; in fact much of the laborious work still remained. They required filling with water, cold or previously heated, and emptying manually; constant attention was needed all through the various parts of the operation, to heat the water but prevent it boiling over, to add the soap or power, to untangle clothes if they got twisted round the agitator, which happened frequently, to remove the clothes for rinsing, and to wring them out. In addition the early machines often leaked, went rusty, did not boil satisfactorily, and some even burst into flames due to faulty construction. In general women seemed to prefer to send out their wash to a laundry or washerwoman if one could be found. Elsie's family, for example, were in great demand:

> My gran had a good reputation and she always had people waiting
> to be customers, sometimes she would take their washing and give it
> out to other women. There were always a lot of women who wanted
> to earn money this way, so why not spread it around a bit she said.

Elsie's grandmother's customers mostly came from the middle classes, 'people who lived in the big houses round the park', but she also washed for the local shopkeepers who had no time to do their own washing, a local woman whose husband worked in London (and therefore earned more money), women who kept boarding-houses, or who rented out rooms to holidaymakers, and some elderly women who could no longer manage it themselves.

Audrey, in common with other women in my sample, recalled that her mother took advantage of the local laundry services whenever she could afford it, especially for bigger items like sheets. This is borne out by evidence from a number of Mass Observation surveys, which show that working-class households did use commercial laundries, especially the cheaper services such as the bag wash. Some women had to use these services because they simply did not have the basic facilities for washing at home but would certainly not have been able to afford a washing machine, nor indeed have had room to house it.[11] The Women's Group on Public Welfare said that over 80 per cent of the poor population of London used the bag wash.[12]

Other reasons for the slow take-up of washing machines at this time can be attributed to the lack of standardization of electricity systems and the consequent inability of manufacturers to adopt mass production methods; and electricity was still not available to everyone.[13] There were still numerous different AC and DC systems in use in the late 1930s, sometimes even within the same town or area.[14] Even wealthy customers were reluctant to invest in expensive appliances which might be useless if they moved only a few streets away. Manufacturers had to make different models for different voltages which kept supplies low and prices high. Moreover, it was often difficult to obtain a washing machine at all; few were in stock in shops which did sell them and marketing was by no means aggressive.[15]

Finally, temperament may have had something to do with it. Certainly for working-class women the sheer struggle of day-to-day living left them with very little energy for envisaging anything other than coping as best they could. Even those in the slightly better-off groups of the working classes did not expect anything but hard work. In response to a question about whether her mother complained about the hard work of washday, Joan, another of my informants, said:

> No. No, I think women accepted all that heavy work. I know I did when I got married in 1940 and there weren't any washing machines and things. I think you just accepted it, that it was part of women's chores and you just got on with it.

Eva, another woman I interviewed, said: 'Well dear you didn't expect anything, you just got on with your work, someone had to do it and you didn't have time to think about it'. These were quite typical attitudes displayed by several of the women I interviewed who in fact seemed to think it was a strange question as the work was so inevitable and 'their' job.

However, by the mid 1930s a little innovation was creeping into some working-class homes. 'New' technologies commonly mentioned by the women I interviewed included soap powder, the Acme wringer, gas and electric boilers and the electric iron. Audrey's great-grandmother, for example, would only use hard soap as she had always done, but Audrey's mother was modern and tried out the new soap powders.

> Oh, my great-grandmother didn't believe in soap powders. My mother used to use Rinso and Oxydol, she had packets of that, but grandma always had bars of Sunlight soap and a scrubbing brush and she wouldn't use any other soap but Sunlight. She said washing powders didn't do the job properly. But my mother wasn't very domesticated and she was willing to try all the new things.

The Acme wringer was mentioned by many as a major new acquisition. Audrey's mother obtained the wringer by weekly payments from a man who called round selling them. Elsie's mother bought one second-hand. Gas or electric boilers were also seen as a great improvement by several women. Audrey talked about the thrill experienced by the women in her home when they got one and likened it to having a de luxe washing machine nowadays after using an old twin-tub.

> People didn't have to be very rich to get one because no one ever bought gas cookers or gas coppers for cash, you used to pay a very little every week, for evermore, they were hired. Just a few coppers a week it was.

Edith, however, told me that she did have an electric copper for a while, but went back to using her old copper as she could not afford the hire charges as well as the extra electricity.

The electric iron was the most important new piece of equipment. Olive's mother hired an iron together with her electric boiler and an electric kettle in 1931 when they moved to a new council house, with electricity. In 1935 only 33 per cent of households had an electric iron. However, this rapidly changed: by 1939 three-quarters of those with electricity had an electric iron, some six or seven million had been produced in the intervening few years and at prices more widely affordable, from 5s to 15s for a standard non-thermostatic iron.

The Electrical Association for Women[16] published an optimistic report in 1935[17] showing that working-class women spent much less time on specific household tasks once their homes had been electrified.

> The average housewife in an unwired home spent just over $26\frac{1}{4}$ hours a week attending to lamps, cooking stoves, and fires, washing, ironing and cleaning, while in a house lit and powered by electricity her work load was reduced by 73 per cent to just over 7 hours.[18]

The report showed that some tasks such as cleaning and filling the lamps, tending ranges and making fires could disappear altogether, and it claimed that time spent on other jobs such as ironing went from 2.95 hours to 1.55; washing, 4.03 hours to 1.60; and cleaning from 8.26 to 4.12 hours.[19] Of course such dramatic changes depended upon being able to acquire all the appliances. The chances of having them were, for most working-class women, still very limited at this time.

However, in principle there was a chance that the unremitting drudgery of heavy housework might begin to be alleviated through the use of 'new' modern technologies and electrical appliances. Together with other changes such as smaller families, which, as Diana Gittins indicates, were becoming much more widespread,[20] household tasks might be lightened and the time spent shortened. However, at this same period the *ideology* of domesticity, focusing on a woman's place as ever-caring wife and mother, was being elaborated through the consumption-oriented women's magazines. At first these were aimed at middle-class women but they were soon followed by magazines specifically for working-class women encouraging them to try out new goods, and indeed training their women readers for consumption in both advertisements and articles.[21] An ideological prescription that women were responsible for the total well-being, not just the physical care, of their family began to take root, so that by the 1950s cooking and cleaning had become invested with love and guilt. Women were to become trapped by the demands of emotionalized housework for many years to come, albeit with 'new' technologies with which to do the work.[22]

In conclusion, taking the period 1914–1945 as a whole it can be seen that there was relatively little change in how women did the washing – and most other housework tasks in working-class homes – but the conditions for change, both social and technological, were gradually being put in place.

Notes

1   See S. Berger Gluck and D. Patai (Eds) (1991), *Women's Words: The Feminist Practice of Oral History*, New York, Routledge; L. Stanley (Ed.) (1990), *Feminist Praxis: Research Theory and Epistemology in Feminist Sociology*, London, Routledge; P. Thompson (1988), *The Voice of the Past*, London, Oxford University Press; T. Lummis (1987), *Listening to History: The Authenticity of Oral Evidence*, London, Hutchinson.
2   C. Zmroczek (1992), 'Dirty Linen: Women, Class and Washing Machines 1920s–1960s', *Women's Studies International Forum*, **15** (2), pp. 173–85; C. Davidson (1982), *A Women's Work is Never Done: A History of Housework in the British Isles 1650–1950*, London, Chatto and Windus, P. 160.
3   M. Spring Rice (1981), Working-Class Wives: Their Health and Conditions, London, Virago (first published by Penguin Books, 1939).
4   L. Hannah (1979), *Electricity before Nationalization: A Study of the Development of the Electricity Supply Industry in Britain to 1948*, London, Macmillan.
5   The Heating and Ventilation (Reconstruction) Committee of the Building Research Board of the Department of Scientific and Industrial Research (1945), *Heating and Ventilation of Buildings*, Postwar Buildings Series, 19.
6   J.C. Wilson (1949), *An Enquiry into Communal Laundry Facilities*, National Buildings Studies Special Report No. 9, HMSO.
7   Public wash-houses were not available in all towns; most were in London, the North of England, and Scotland. See Davidson (1982), p. 163, and Wilson (1949). They were not always popular; see, for example, Mass Observation (1939), *Clothes Washing, Motives and Methods: Interim Report* (MO reference: A18), Mass Observation Archive, University of Sussex.
8   R. Schwartz Cowan (1983), *More Work for Mother*, New York, Basic Books.
9   See Spring Rice (1981).
10 'Electricity is the Servant of all Housewives', *Ideal Home*, Feb. 1930, p. 106, in an article on 'Aids of Electric Power', including cookers, fires, vacuum cleaners, floor polishers, water heaters, toasters, and percolators. See also A. Forty (1986), *Objects of Desire: Design and Society 1750–1980*, London, Thames and Hudson, pp. 211–15.
11 Mass Observation (1939), *Clothes Washing: Motives and Methods: Interim Report*; *Laundry Usage (1945)*, Survey prepared for the Institute of British Launderers, Ltd, by the Research Department, London Press Exchange Ltd, January 1946 (MO Reference: 2315), Mass Observation Archive, University of Sussex; W.F.F. Kemsley and D. Ginsburg (1949 (August)), *Expenditure on Laundries, Dyeing and Cleaning, Mending and Alterations and Shoe Repairing Services*, The Social Survey Consumer Expenditures Series, Central Office of Information.

12 Women's Group on Public Welfare (1943), *Our Towns: A Close Up*, London, Oxford University Press.

13 In 1935 only 54 per cent of homes were wired for electricity. By 1938 this had risen to over two-thirds of all housing (Hannah, 1979, p. 188).

14 Hannah (1979), p. 196.

15 See Mass Observation (1949) Shopping Box 5HA *Report on a Washing Machine* (sixty interviews with retailers), Mass Observation Archive, University of Sussex; and *Electrical Review* **XCVII** (1925), pp. 366–8, about attempts to purchase a washing machine.

16 For more information on the Association, see S. Worden, 'Powerful Women: Electricity in the Home 1919–1940', in J. Attfield and P. Kirkham (1989), *A View from the Interior: Feminism, Women and Design*, London, The Women's Press, pp. 131–50.

17 E.E. Edwards (1935), *Report on Electricity in Working Class Homes*, London, Electrical Association for Women.

18 Davidson (1982), p. 43.

19 *Ibid.*, p. 43.

20 D. Gittins (1982), *Fair Sex: Family Size and Structure 1900–39*, London, Hutchinson, p. 31. Gittins says that 67 per cent of the population married in 1925 had two or less children. However, most of the women in my study came from large families; the overall average was four children per family. The same average is confirmed by Margery Spring Rice who comments in her survey of working-class women on the lack of evidence for the decline of family size.

21 See, for example, D. Beddoe (1989), *Back to Home and Duty*, London, Pandora, ch. 1.

22 For more discussion of the changing shape of household demands see R. Schwartz Cowan, 'The "Industrial Revolution" in the Home: Household Technology and Social Change in the 20th Century', in M. Moore Trescott (1979), *Dynamos and Virgins Revisited: Women and Technological Change in History*, London, Scarecrow Press; R. Schwartz Cowan, 'A Case Study of Technological and Social Change: The Washing Machine and the Working Wife', in M.S. Hartmann and L. Banner (Eds) (1974), *Clio's Consciousness Raised: New Perspectives on the History of Women*, London, Harper and Row; S. Strasser, (1992), *Never Done: A History of American Housework*, New York, Pantheon Books; S. Jackson (1992), 'Towards a Historical Sociology of Housework: A Materialist Feminist Analysis', *Women's Studies International Forum*, **15** (2), pp. 153–72.

Chapter 2

A 'Trade Union for Married Women': The Women's Co-operative Guild 1914–1920

Gillian Scott

The 1910s were years of unprecedented change for women in Britain. Organized feminism, on the one hand, and the upheavals induced by war, on the other, produced a range of cultural shifts and political and social reforms which constitute a watershed in women's history. In this chapter I concentrate on the character and experience of what was arguably the most outstanding women's organization of this period – the Women's Co-operative Guild.

The WCG – an auxiliary body of the consumers' co-operative movement – was founded in 1883; during the 1890s and early 1900s its original, and relatively modest, aim of teaching women about co-operation was transformed into a far more political agenda which articulated the needs of a previously unrepresented constituency – working-class wives and mothers. 'Being composed of married women who are co-operators', explained the General Secretary, Margaret Llewelyn Davies, in 1920, 'it has naturally become a sort of trade union for married women'.[1] With 30,000 members in 1914, and 50,000 in 1921, predominantly working-class housewives who shopped at the co-operative stores, its great achievement had been to orchestrate 'the emergence of the married working-woman from national obscurity into a position of national importance'.[2]

The Guild's distinctive reputation was based on its pioneering work in two areas: divorce law reform and maternity care. These initiatives opened up a perspective on the rights and wrongs of women previously excluded from political discourse. Since the mid-nineteenth century organized feminism had sought to increase women's access to the public sphere, but had been relatively silent about the private sphere.[3] While the Guild also supported women's equal participation in political and economic life – the co-operative store, the town council and Parliament – it recognized that exclusion from public life was only part of the problem. As well as seeing her relationship with public activity, explained one Guild publication, the married working woman 'reflects on her own position in her home, and ... understands that the first need is equal comradeship of husband and wife, of brothers and sisters'.[4]

In very many cases conditions in the home were anything but equal.

For generations, the very concept of the 'private' sphere facilitated the concealment of family life from public gaze. 'In the past', wrote Davies, 'a heavy curtain had, on marriage, fallen on the woman's life, for the nation felt no responsibility for her personal welfare or for the conditions under which she performed her great tasks'.[5] This curtain was weighed down with domestic ideology, 'embroidered' with 'all sorts of beautiful sentiments about the beauty of motherhood and the sanctity of the home'.[6] Yet the reality was often considerably less than beautiful.

> Without money of her own, with no right even to her husband's savings, without adequate protection against a husband's possible cruelty, with no legal position as a mother, with the conditions of maternity totally neglected, married women in the home had existed apart, voiceless and unseen.[7]

These circumstances endured while the 'isolation of women in married life ... prevented any common expression of their needs',[8] but the Guild's development began to change this atomized situation. Organization enabled women to learn that their experience was not unique: 'now the curtain was being withdrawn, and from the discussions that had taken place they had learned much of the sufferings of married women, the pain and misery that were going on behind the curtain'.[9] It is this focus on women's situation within the private sphere of the working-class family – its insistence on the urgency of 'the need for reforms in the lives of the married women themselves'[10] that sets the Guild apart from other feminist organizations of the time and makes its contribution to social progress and the cause of women so important.

The Guild's achievements during this period are in no small measure related to the organizational and political leadership of Margaret Llewelyn Davies. One of her guiding principles as General Secretary was that 'so long as there is class and sex inequality, it is necessary that working women should have their own separate and affiliated organisations'.[11] This was not an easy position to hold on to. Working at 'the hyphen'[12] of class and gender politics meant that the Guild could not assume the official support of the working-class movement of which it was a part; to the contrary, it encountered hostility to its policies on a number of occasions, and, from 1914 to 1918, was in dispute with the co-operative leadership over the related issues of divorce law reform and self-government. The Guild's part in this conflict illuminates not only its approach to the question of women's subordination in the home but also the clarity and vigour of the members' defence of an autonomous, working-class, women's politics.

Divorce Law

'You will see from *Co-operative News*', Davies wrote to Leonard Woolf in April 1914, 'that we are in a nice hole as regards Divorce'.[13] The Guild's support for radical reform of the divorce law had attracted the disapprobation of the Manchester and Salford Catholic Federation which maintained that Catholic members would withdraw from the Societies in their thousands if the Guild did not drop the issue.[14] This largely unfounded threat got short shrift from the WCG Central Committee – they 'considered the reform of divorce law to be one of the most important moral and social reforms which affect co-operative women'[15] – but found greater resonance with the executive body of the Co-operative Union – the Central Board. For many years the Board had objected to the Guild's 'Citizenship' work on women's rights, and they now leapt at the opportunity to curb the organization's freedom of action. On the grounds that 'If the board had to pay the piper, they should be able to call the tune',[16] it was ruled that the Guild's £400 annual grant would be paid on the condition that 'they cease agitation in favour of the alteration of the divorce law. That in future the Women's Guild be requested not to take up any work disapproved of by the Board'.[17]

Yet far from bringing the Guild to heel, the 'guardians of the prestige of the movement'[18] succeeded only in eliciting a demonstration of the organization's strength and unity in relation to the issue of divorce law reform in particular and that of its own self-government in general. At Congress that year the 864 delegates passed, without opposition, a resolution stating that 'the future progress of the Guild and of the Co-operative Movement depends on the guild policy being democratically controlled by the members themselves'.[19] This position was defended in a series of powerful contributions. One woman exclaimed at 'the ignorance of the union. Did they think that they were the only ones who knew what was good for the women. Then they forgot that the women could think for themselves. (Applause)'. Another claimed that she knew 'what the proposed reforms meant for downtrodden women, and she could not help but raise her voice in protest of the action taken by the Co-operative Union. . . . It was to be regretted that there were such men in the movement'. Guildswomen, another delegate explained,

> wanted to work with the men side by side, not as subordinates with restrictions, for they possessed the powers and abilities of adult women. (Hear, hear, and applause.) They were not prepared to be dominated. (Loud and prolonged applause.) They were open to criticism; but they did not take any action without having first carefully considered the question. (Hear, hear.) . . . This subject of Divorce Law they had been considering for the past four years; but for how many years had women been suffering?[20]

As these spirited interventions suggest, the outrage generated among guildswomen at the suggestion of bureaucratic interference was in direct proportion to the sense of grievance generated by the Guild's investigation of the divorce issue. In 1910, when invited to submit evidence to the Royal Commission on Divorce and Matrimonial Causes, the Central Committee carried out a survey – presumably the first of its kind – to establish the views of working women on the subject, and it was this inquiry that first 'brought to light' the 'hidden suffering' within many working-class marriages.[21] The substance of the Guild's submission to the Commission came from the membership in the form of 131 'manuscript letters – often many pages long, laboriously written after thought and consultation',[22] describing broken marriages in which, as the General Secretary noted, 'the suffering is evidently much more on the side of women than on the side of men. . . . No woman could inflict on a man the amount of degradation that a man may force on a woman'.[23]

Ignoring the convention that what went on in the home was a private matter between husband and wife, the Guild's evidence disclosed the seamier side of married life. Cases were cited 'where a woman, ill-used and kicked, has taken her husband back five times; of a diseased husband compelling co-habitation, regardless of his wife's health; of a man frightening his wife during pregnancy in order to bring on a miscarriage'.[24] One woman had been 'compelled to submit' to her husband only days after the birth of her baby; after her third child, her husband's demands made her so ill that the doctor told her to 'go from home and stay as long as possible'.[25] A guildswoman described women she knew 'who always try to bring on an abortion when first they are pregnant . . . because the husband will grumble and make things unpleasant, because there will be another mouth to fill and he may have to deprive himself of something'.[26]

On the basis of such evidence, the Guild was able to make a strong case for sweeping changes: divorce should be cheaper, the grounds should be equal between men and women, and considerably extended to include incompatibility or mutual consent. In doing so, it went against the grain of public opinion. A number of the recommendations of the liberal Majority Report, published in 1912, were in line with Guild proposals, although they did not go so far as to include incompatibility, but the war intervened and it was not until 1923 that the contentious double standard – which enabled a man to divorce his wife for adultery while a wife must prove some additional ground – was abolished, while the grounds for divorce were not extended until 1937. The Guild's proposal that mutual consent be a legitimate reason for divorce did not become statutory until 1969, a time lag that provides a barometer of the climate of thought on the subject half a century previously.[27]

Divorce was a taboo subject. The Central Board recognized that it was highly controversial and believed that the Guild's involvement with the issue would bring the co-operative movement into disrepute. The Church basically considered it to be a sin. The Minority Report of the Royal Commission, signed by an archbishop and two ecclesiastical lawyers,[28] claimed that the Guild 'advocated a facility of divorce hitherto unheard of in any civilised country'.[29] They considered marriage to be 'a lifelong obligation with all the sacrifices which such an obligation involves'.[30] From a rather different perspective, a number of feminists maintained that divorce was a loose cannon that could be turned against women. Millicent Fawcett in her evidence to the Commission 'opposed any extension of the grounds for divorce other than for permanent separation or wilful desertion'. While such liberal feminists supported ' "levelling up" in order to raise men's moral standards', they upheld the marriage contract as a necessary mechanism for guaranteeing the economic security of women and children and opposed any reform that would make it easier for men to escape their responsibilities.[31]

Why, then, should a working women's organization, the majority of whose members were economically dependent upon their husbands, and embedded in the conformism of the respectable working class, be so unequivocally in favour of reform? First, it is significant that the Guild sought to foster confidence and self-esteem in its members and explicitly rejected the traditional valorization of womanly self-sacrifice. Davies, for example, pointed out to the Commissioners: 'If divorce is considered a sin, and the patient endurance of degradation and compulsory suffering a virtue, a most serious moral confusion is created. It means that women's self-respect and happiness are sacrificed'.[32] Some Guildswomen were said to be 'diffident about discussing their home affairs'[33] and shrank from the subject of divorce, but they were reminded of 'the responsibility that is laid upon the Guild, as an organisation of married women, to take the lead in a campaign which is concerned with the dignity and respect of married women'.[34] The level of debate at Congress demonstrates the Guild's success in empowering its members to do precisely that.

Secondly, and relatedly, the Guild did not pretend that legal changes alone would solve all the problems that were highlighted in its evidence. Davies owned that 'Even if unjust laws and public opinion were changed, the greater physical weakness, the conditions of maternity, the difficulty of monetary independence, would still put power in the hands of men, and give opportunities for its abuse'. She also acknowledged the argument that the marriage tie should be made indissoluble 'for the sake of the women themselves and their children'. She admitted that the economic problem produced 'an extremely difficult complication'[35] that 'would have to be dealt with', but she insisted that 'our attitude to divorce should not depend upon it'. To the contrary, she argued that to allow the economic factors to determine the legal relationship was to tempt women to 'sacrifice their personal dignity and honour' for 'what is little better than prostitution'.[36]

Critically, the Guild did not regard divorce law reform as the only means by which the position of working women might be advanced, any more than it viewed the vote as an end in itself rather than an instrument for securing further change. As part of the co-operative movement and as a campaigning organization with its own agenda, the Guild pursued equality for women as part of a wider struggle for 'a social system founded on fellowship ... a real industrial and political democracy' which would abolish 'class distinctions'.[37] Thus legal and political rights were seen in relation to the social reforms that were also needed by working women, to give substance to formal entitlements, not as minor modifications of an otherwise immutable social order.

As the material on divorce law made plain, the pivotal difficulty for married women was their economic dependence. Many cases were cited of husbands not giving 'sufficient money for the maintenance of the family';[38] and sympathy expressed for women's legal entitlement to a definite share of the wages: 'Men have the idea a woman's work in the home is not work, but it is work that should be paid for just the same as any other'.[39] Yet the Guild also recognized that buttressing the unequal division of the family income was the inadequacy of wage levels. In general terms, the Guild supported both trade union and co-operative strategies to improve working-class living standards, but more immediately it set out to meet the needs of women by pushing for the state to assume some of the cost of reproduction, in the first instance by instituting a scheme for the 'national care of maternity'.

Maternity

While the outbreak of war brought an adjournment of the campaign for divorce law reform, the Guild's work for statutory maternity provision received a boost from the sudden concern for the health and well-being of working-class mothers engendered by the ruthless slaughter of their sons on the battlefields. The Guild's credentials in this area had already been established through its efforts to ensure that maternity benefit was included in the 1911 National Insurance Act, and then in the passage of the 1913 amending bill to make it the legal property of the mother. The WCG 'citizenship sub-committee' had collected case studies from guildswomen, organized petitions, deputations, and letters to the press and to MPs, attended the relevant hearings in the House of Commons, and eventually won a resounding victory: 'the first public recognition of the mother's place in the home, and the first step towards some economic independence for wives'.[40]

It is interesting to note that in pressing for state welfare payments, the Guild encountered a form of opposition which presaged its subsequent differences with the Co-operative Union and underlined the salience of its claim that 'a body practically composed of men does not understand or give due consideration to the views of women'.[41] At the committee stage of the 1913 Bill, the 'chief opposition' to the Guild's position came from 'the five Labour men' who 'took a definite line against the view' that the benefit should belong to the wife. Again, in the House of Commons, the Labour MPs were explicitly against the measure. As the Women's Pages of *Co-operative News* reported despairingly: 'what can be said of the position taken up by Mr Roberts (the Labour member), who argued that to insist on the money being paid to the woman was an insult to the working man?'[42] Underlying this objection, no doubt, was the concern, regularly voiced by trade-union leaders, that state welfare payments would weaken their position in collective bargaining, especially in regard to their claim that they worked for a 'family wage'.[43]

Davies' introduction to the Guild's 1915 publication, *Maternity: Letters from Working Women*, emphasized that 'the conditions of life which our industrial system forces upon the wage earners' was a root cause of the 'perpetual overwork, illness and suffering' described in the 160 letters. She drew a telling comparison between the conditions of maternity experienced by women of different social classes:

> The middle-class wife from the first moment is within reach of medical advice which can alleviate distressing illness and confinements and often prevent future ill-health or death. During the months of pregnancy she is not called upon to work; she is well fed; she is able to take the necessary rest and exercise. At the time of the birth she will have the constant attendance of doctor and nurse, and she will remain in bed until she is well enough to get up. For a woman of the middle class to be deprived of any one of these things would be considered an outrage. Now, a working-class woman is habitually deprived of them all.[44]

Yet Davies was not claiming that these women's problems could simply be reduced to the material circumstances of the working-class family. Alongside inadequate wages, she also drew attention to the 'delicate' matter of the 'personal' relation of 'husband and wife':

> In plain language, both in law and in popular morality, the wife is still the inferior in the family to the husband. She is first without economic independence, and the law therefore gives the man, whether he be good or bad, a terrible power over her ... the beginning and end of the working woman's life and duty is still regarded by many as the care of the household, the satisfaction of man's desires, and the bearing of children.[45]

As with the material collected for the Divorce Commission, the *Maternity* letters brought out aspects of women's sexual oppression that were rarely, if ever, discussed in public. 'I often think women are really worse off than beasts', wrote a woman who had had seven children and two miscarriages.

> During the time of pregnancy, the male beast keeps entirely away from the female: not so with the woman; she is at the prey of a man just the same as though she was not pregnant. Practically within a few days of the birth, and as soon as the birth is over, she is tortured again. If the woman does not feel well she must not say so, as a man has such a lot of ways of punishing a woman if she does not give in to him.[46]

Reflection on their experience of maternity prompted many women to discuss various methods of birth control. Some women spoke gratefully of 'considerate' husbands, while the use of abortifacients was discussed matter-of-factly by a number of respondents. 'Can we wonder', reflected one, 'that so many women take drugs, hoping to get rid of the expected child?'[47] A small number of women referred to 'Preventitives'; one woman, for example, described how, after bearing seven children in ten years, she had decided that 'if there were no natural means of prevention, then, of course, artificial means must be employed, which were successful'.[48]

While the Guild's primary concern in publishing these letters was to highlight the appalling conditions of maternity for many women, the decision not to edit out the controversial references to contraception and illegal abortion indicates a determination to emphasize the necessary connections between the two issues. Davies, in her introduction, claimed that the steady decline in the birth rate was 'mainly due to the deliberate limitation of the family'. Whether one considered this 'kind of strike against large families' to be 'the suicide of a nation and the doom of a race' or the 'clearest solution of the inextricable tangle in which the industrial system has enmeshed humanity',[49] was largely irrelevant, given the strong evidence to show that if 'maternity is only followed by an addition to the daily life of suffering, want, overwork, and poverty, people will continue to adopt even the most dangerous, uncertain, and disastrous methods of avoiding it.'[50]

The implication of Davies' argument is that it would be to the benefit of working men and women alike for safe and reliable methods of birth control to be made easily available. Clearly this was not the time or the place to initiate such a campaign, but the general tenor of the Guild's treatment of divorce law and maternity during this period suggests that, as soon as a suitable opportunity arrived, it would be to the fore in advancing working women's claims in this area.

Later History

In many respects the end of the war saw the Guild in a stronger position than at any time in its history. The enfranchisement of women over 30, and the creation of the Co-operative Party, improved its bargaining position with the Co-operative Union. A settlement was reached, the grant restored, and the Guild, which had renewed its work for divorce law reform, re-stated its conviction that a self-governing women's auxiliary body was the best means of 'raising women to their proper status in the movement . . . and securing attention for the reforms needed by them as wives, mothers and housewives'.[51] A new mood of confidence among co-operative women was apparent in the Guild's rapid growth – 11,000 new members from 1919 to 1920 and another 6,000 in the following year, bringing its numbers to just over 50,000 by 1921.

Despite these encouraging developments, during the 1920s there was a discernible blunting of what had been the radical edge of Guild politics. In certain areas it maintained the continuities of the previous decade. *Maternity* had included the Guild's proposals for a comprehensive national scheme for the care of maternity to be administered by the Public Health Authorities. Certain of their demands were reflected in the Local Government Board directives issued during the war, others found expression in the 1918 Maternal and Infant Welfare Act, but a number – such as the provision of maternity homes for normal births – took decades to achieve and remained a staple of Guild campaigns, locally and nationally, during the inter-war years, placed now in the wider context of the demand for the 'State Endowment of Mothers and Children', also known as Family Endowment and later as Family Allowances.[52]

Increasingly, however, the Guild concentrated on 'safe' women's issues, such as maternity care, and neglected the controversial subjects that had previously been lifted out of the private sphere and into the public domain. In the early 1920s, as the climate surrounding birth control began to shift, many guildswomen supported the demand for greater access to contraceptive advice through the new maternity clinics. When a resolution to this effect was passed at Congress in 1923, it drew fire from the Guild's old adversary – the Manchester and Salford Catholic Federation. The Central Committee's response this time around was to decide not to make birth control a special subject, but to advise branches who were interested to take up the matter locally.[53]

Following Margaret Llewelyn Davies' retirement in 1921, a new generation of Guild leaders steadily shifted away from the earlier principle of an autonomous working women's movement, and hitched their wagon to the engine of Co-operative and Labour Party politics. One of the consequences of this development was a growing tendency for Guild policy to reflect the leadership's concern not to take up sensitive or controversial subjects, expecially not 'sex questions' that might be divisive. In this the WCG kept step with the official line of the Labour Party Women's Section, with which it worked closely at a national level on the Standing Joint Committee of Working Women's Organisations. We 'do plead', ran an editorial in the *The Labour Woman*, on the subject of contraception 'that this subject of the relations of husband and wife should not be treated as a political issue at all'.[54]

The changes that took place in the WCG during the 1920s demonstrate that the distinctive politics which characterized it a decade earlier did not arise 'naturally' from its social base among co-operative women, but were produced by historical and political circumstances which were both specific and contingent. After the war, the new leadership, the decline of organized feminism, and realignments in the working-class movement, all combined to pull the Guild away from its previous independent and innovatory approach to politics, and towards a more anodyne and exclusively parliamentary set of concerns in which it was generally overshadowed by the Labour Party Women's Sections.[55]

The concept of 'sexual politics' did not surface again until the growth of the Women's Liberation Movement in the 1970s when it was reworked under the slogan 'The Personal is Political'. Despite some of the similar foci of the Guild and this 'second-wave' feminist movement (e.g. divorce law reform and defence of abortion rights), there are also significant differences. The radical feminist redefinition of politics as something that 'could take place even in the intimacy of one's bedroom'[56] constituted an inversion of the Guild's approach, in identifying individual men, as opposed to juridical and structural aspects of society, as the source of women's oppression. Furthermore, despite certain efforts in that direction, 'second-wave' feminism was unable to achieve the large-scale, working-class base which had made the early WCG such a powerful force in the co-operative movement and beyond.

Notes

1 M.L. Davies, 'Co-operation at the Fountainhead', *Life and Labour*, Chicago, vol. K, no. 7 (Sept. 1920), pp. 199–202, typed MS, 'Material Illustrating the Work of the Guild and Kindred Interests . . . Erstwhile Property of M. Llewelyn Davies (1890–?1944)', 11 vols, British Library of Political and Economic Science, vol. 1, item 25.
2 M.L. Davies, 'Foreword', in C. Webb (1927), *The Woman with the Basket*, Manchester, Women's Co-operative Guild.
3 J. Lewis (1984), *Women in England 1870–1950: Sexual Divisions and Social Change*, Brighton, Wheatsheaf, p. 103.
4 Women's Co-operative Guild (1913), *The Education of Guildswomen*, p. 4.
5 Davies, article for *Norges Kvinder*, Norwegian women's paper (1931), typed MS, 'Material . . .' (see note 1), vol. 1, item 39.
6 *Co-operative News*, 27 May 1911, p. 667.
7 Davies, 'Foreword', in C. Webb (1927).
8 M.L. Davies, 'Introduction', in M. Llewelyn Davies (Ed.), *Maternity: Letters from Working Women* (Virago, 1978, first published 1915), pp. 8–9.
9 *Co-operative News*, 27 May 1911, p. 667.
10 'The WCG, 1895–1916', *Annual Congress Handbook* (London, 1916), p. 11.

11 'Report of the English Women's Co-operative Guild' (1921), typed MS, 'Material ...' (see note 1), vol. 1, item 30.
12 See Ellen DuBois, 'Women Suffrage and the Left: An International Socialist Feminist Perspective', *New Left Review*, 186 (March/April 1991) for a discussion of woman activists at the turn of the century working at 'the hyphen' between socialist parties and the women's movement.
13 M. Llewelyn Davies to Leonard Woolf, n.d. (? March 1914), Monks House Papers, University of Sussex.
14 Manchester and Salford Catholic Federation to the WCG Central Committee, Oct. 1913, Monks House Papers, University of Sussex, MS 18.
15 *Co-operative News*, 28 March 1914, p. 398.
16 Central Board Meeting, *Co-operative Union Annual Report*, 1914, p. 15.
17 *CUAR*, 1914, p. 574.
18 *CUAR*, 1914, p. 1.
19 WCG, *Annual Report*, 1914–15, p. 5.
20 *Co-operative News*, 20 June 1914, p. 808.
21 'The WCG, 1895–1916', p. 11.
22 'Evidence of Miss Margaret Llewelyn Davies', *Minutes of Evidence Taken before the Royal Commission on Divorce and Matrimonial Causes* (1912), vol. 3 [Cd. 6481], PP 1912–13, XX, p. 149.
23 'Evidence of Miss Davies', *op. cit.*, pp. 150–1.
24 Women's Co-operative Guild (1913), *Divorce Law Reform: The Majority Report of the Divorce Commission*, Manchester, Women's Co-operative Guild, p. 12.
25 'Evidence of Miss Davies', *op. cit.*, p. 166.
26 *Ibid.*, p. 154.
27 *Report of the Royal Commission on Divorce and Matrimonial Causes* (1912) [Cd. 6478], PP 1912–13, XVIII; see J. Weeks (1989), *Sex Politics and Society: The Regulation of Sexuality since 1800*, Harlow, Longman, for a useful account of changes in the divorce laws, especially ch. 13, 'The Permissive Moment', for the context of the 1969 Divorce Act.
28 *Co-operative News*, 19 April 1913, p. 499.
29 Minority Report, *Report of the RC on Divorce*, p. 177.
30 *Ibid.*, p. 188.
31 Lewis (1984), p. 130.
32 'Evidence of Miss Davies', *op. cit.*, p. 162.
33 *Co-operative News*, 19 April 1913, p. 500.
34 *Co-operative News*, 26 April 1913, p. 529.
35 'Evidence of Miss Davies', *op. cit.*, p. 161.
36 *Ibid.*, p. 162.
37 M. Llewelyn Davies, 'Special Education, Divorce and Independence', article for German paper, typed MS, 'Material ...' (see note 1), item 42, ?1933.
38 'Evidence of Miss Davies', *op. cit.*, p. 169.
39 *Ibid.*, p. 172.
40 *Co-operative News*, 23 August 1913, p. 1083.
41 WCG Annual Congress, 1915 *The Self Government of the Guild*, p. 3.
42 *Co-operative News*, 16 August 1913, p. 1038.
43 M. Barrett and M. McIntosh, 'The "Family Wage": Some Problems for Socialists and Feminists', and J. Brenner and M. Ramas, 'Rethinking Women's Oppression', both in T. Lovell (Ed.) (1990), *British Feminist Thought: A Reader*, Oxford, Blackwell, focus a number of the key issues and debates generated by the concept of the 'family wage'.
44 Davies, *Maternity*, pp. 3, 4–5.
45 *Ibid.*, pp. 7–8.
46 *Ibid.*, pp. 48–9.

47   *Ibid.*, p. 25.
48   *Ibid.*, pp. 60–1.
49   *Ibid.*, p. 13.
50   *Ibid.*, p. 15.
51   WCG, *Annual Reports*: 1918–19, p. 4; 1919–20, p. 3.
52   'Annual Report of Women's Guild', in *CUAR*, 1920, p. 277; M.L. Davies, 'The Claims of Mothers and Children', in M. Phillips (Ed.) (n.d.), *Women and the Labour Party*, London, Headley Bros; see also S. Fleming, 'Introductory Essay', in E. Rathbone (1984), *The Disinherited Family*, Bristol, Falling Wall Press, for an account of various strategies to improve the financial position of married women during the 1920s and 1930s.
53   WCG, Central Committee Minutes, 16 July 1923; 19 and 20 Sept. 1923.
54   *The Labour Woman*, XII/2, 1 March 1924, p. 34.
55   The exception to this pattern, the Guild's vigorous support for pacifism, led, in the late 1930s, to an uncomfortable alliance with Conservative appeasement on one side, and ethical and religious pacifists on the other, neither of which had any particular interest in promoting the interests of working-class women.
56   D. Coole (1988), *Women in Political Theory from Ancient Misogyny to Contemporary Feminism*, Brighton, Harvester Wheatsheaf, p. 256.

Chapter 3

The Women's Institute Movement –
The Acceptable Face of Feminism?

Maggie Morgan

Edith Rigby was born in 1873, the daughter of a Preston doctor. Her suffragette activities, which included planting a bomb in the Liverpool Corn Exchange, pouring acid on the green of a local golf course, and setting fire to the stands of Blackburn Rovers' football club, resulted in her being a guest at His Majesty's pleasure some seven times. During her imprisonment she went on hunger strike and consequently suffered forced feeding. After setting fire to Lord Leverhulme's house at Rivington Pike in 1913 she declared from the dock:

> I want to ask Sir William Leverhulme whether he thinks his property on Rivington Pike is more valuable as one of his superfluous houses occasionally opened to people, or as a beacon lighted to King and Country to see here are some intolerable grievances for women.[1]

Edith was a committed socialist, entertaining Keir Hardie amongst others. A contemporary claimed of her:

> Edith was very critical of her neighbours in Winkley Square where she lived with her doctor husband, Charles. They confined their servants to the attics or basements during non-working hours. Her own maids had the run of the house, eating in the dining room, having the evenings free, and did not wear uniforms which Edith considered badges of servitude.[2]

When the First World War started Edith bought a cottage with two acres just outside Preston and began doing her bit to increase the nation's food production. By the 1920s she was a founder member and President of Hutton and Howith Women's Institute (WI) in Lancashire.

The Women's Institute movement was formed in 1915 with funding first from the Agricultural Organization Society and then from the Board of Agriculture. By 1919 the National Federation of Women's Institutes was an independent, democratic organization, publishing its own monthly magazine, *Home and Country*. By 1925 it had a quarter of a million members, a figure it has never fallen below, making it the largest post-suffrage women's organization in Britain. Its function was to improve the lives of rural women and it was non-party-political and non-denominational in character. Edith Rigby's progression from suffrage work to the WI was by no means unique. Virginia Woolf, Elizabeth Robins, author of *Votes for Women*,[3] and Mrs Auerbach, treasurer of the NUWSS, were all involved in the WI in the 1920s and 1930s. To women of the period it was a natural outlet for their feminist activities, as Ray Strachey pointed out in *The Cause*:

> It is not too much to say that the lives of country women were transformed by the coming of this organization, which brought instruction and variety just at the moment when enfranchisement and short skirts were bringing physical and mental development; and it is not surprising that women of all ages and classes who had worked in the suffrage movement turned their energies to this field.[4]

To understand why this organization, founded in 1915, and now associated more with tweeds and twinsets than feminism, was seen as a natural continuation of suffrage work, I want to explore the following areas: the role of the WI in supplying a female cultural space within villages in the inter-war period and their attitudes to domesticity, followed by their activities to improve the standard of rural housing and water supplies. It is necessary, first, to raise questions around the issue of what constitutes a feminist organization and why I feel able to contend that in the inter-war period the NFWI was able to be the acceptable face of feminism in Britain.

The Women's Institute Movement and Feminism

In arguing for the WI to be perceived as feminist I am arguing from my own perception of feminism today. I regard as feminist any activities by groups of women which challenge the boundaries of socially constructed acceptable behaviour for women, whether in economic, political, or cultural terms. Feminism has always been fragmented, at once linked and divided, held together by networks and very loose commonalities. It is vital that the WI be seen a a continuum with the more overtly politicized women's movements both before and in parallel to it. The National Federation of Women's Institutes (NFWI) co-operated with the Six Point Group, the Society for Equal Citizenship, the Women's International League for Peace and Freedom, and the National Council for Women. Equal rights issues such as the campaigns for women police and equal pay, and demands for improved maternity services (a classic maternal feminist campaign) were all on the organization's agenda. In terms of personnel issues and cultural codification there were networks and links going on between a wide variety of women's movements in the inter-war period. Women tended to move from one organization to another, and to prioritize one issue or group of issues according to their political, economic, and cultural specificity at a particular point in time. The WI's ability to be a popular movement relied upon it being able to serve different needs for different groups and to speak at different times for women whose interests were differently influenced by class, age, regionality, marital status, political persuasion, religion, and ethnicity.

The WI was castigated by both the political left and right in the 1920s for its political bias and even by the Church of England. In Yorkshire in the 1920s husbands refused to let their wives join what they described as 'that secret society'. One of the most extreme reactions came in a letter to *Home and Country* in 1928 signed by a 'mere man' and complaining about the amount of time spent on, and involvement with, the WI by wives – the subtext of the letter being presumably this time should be spent in dancing attendance on men. The letter-writer claimed: 'I mix with men, many of whom are husbands of Institute members and the things they say about the Institute are unprintable, one told me this morning "that damned Institute is a curse of a married man's life" '.[5] In order to understand the level of antagonism expressed by the 'mere man', it is necessary to go beyond men's nervousness about female friendship and look at how the WI worked, in the villages, to create for women an alternative cultural space, a form of female-run counter-culture within which they challenged perceptions of the domestic role of rural women.

Female Cultural Space

The Institute in the village operated fundamentally around a monthly meeting backed up by classes as varied as folk dancing or dressmaking, a number of fundraising social events such as whist drives or village socials, and a yearly outing. The meeting consisted of the business, one or two talks or demonstrations, tea (known as the cement of the movement), and the entertainment half hour. A variety of games, dancing, singing, or acting would make up this entertainment. The topics for talks were wide-ranging: Northleigh in Oxfordshire in 1920 covered the use of paper patterns, co-operative buying, local history, where food comes from, haybox cookery, and the Empire, amongst other matters. In 1926 their programme included local government and women voters, sightseeing in Italy, Jerusalem at the present time and the WI movement,[6] whereas Llanfair in 1919 had a talk on Bolshevism in its simplest form.[7] The minutes of Appuldram in Chichester in the 1920s, however, record that the committee did not feel that a proposed talk on VD and public hygiene was necessary.[8]

Getting the regular membership to attend the talks in both mind and body often proved a little difficult. If Funtington in West Sussex had one hundred and fifty people attend a dance it could only muster thirty to a talk on making something out of nothing. It was not alone in this problem, as minute books show. Furthermore the discerning membership (according to records) showed a definite inclination to chat, knit, or snooze through any subject they felt was uninteresting or irrelevant.[9] I point this out not to trivialize the more serious side of the movement, but rather to emphasize that the predominantly working-class members of the Institutes were not powerless: they used the Institute as, when, and how it suited them. If there were problems with ordinary meetings then any of the other suggestions from county or national officers could fare even worse. *Home and Country* frequently ran articles which indicate the leadership's awareness of the different attitudes to WI events. In one entitled 'The Roll Call', a vicar's wife, who was president

of a small Institute, was encouraged by a local WI Voluntary County Organizer (VCO) to have a roll call. It is explained thus – 'the idea is to make all
the members talk and be friendly together so we ask them the same question
and they answer it in turns'. The choice of question, 'What improvement
would you most like in your house?', indicated a wish to raise the membership's
consciousness and precipitate discussion over a key WI concern. The first four
members refused to say anything without more warning, the next four all
opted for electricity and the next two for a wireless. One women felt that her
house would best be knocked down altogether, which was not well received
by her landlady sitting nearby. As the next two women opted for a lodger and
a husband respectively the president decided to unaninmous approval that it
must be tea-time.[10] For these women the importance of the event lay in the
fun and camaraderie of other women rather than in discussing social reforms
whose implementation seemed remote and unlikely.

It was not just the membership for whom the female companionship and
an alternative female culture was important. E.M. Delafield was a WI VCO
and writer who fictionalized her experiences in a regular column for the
feminist magazine *Time and Tide* entitled 'Diary of a Provincial Lady'. She
found herself in a number of situations, judging darning competitions (which
she felt singularly unqualified to do, due to her own incompetence in that
direction) and constantly having her whole household hunt for her membership badge as she left to speak at a local Institute. She undertook a three-day
speaking tour of Institutes which involved one night freezing in a manor
house with a geriatric deaf member of the aristocracy for company. Her
description of her next night is one of female cultural space which will not be
unfamiliar to many women today:

> We talk about the Movement – Annual Meeting at Blackpool per-
> haps a mistake, why not Bristol or Plymouth? – difficulty of thinking
> out new Programmes for monthly meetings, and really magnificent
> performance of Chick at recent Folk-Dancing Rally, at which Insti-
> tute members called upon to go through 'Gathering Peascods' no less
> than three times – two of Chick's best performers, says Assistant
> Secretary proudly, being grandmothers. I express astonished admira-
> tion, and we go on to Village Halls, Sir Oswald Mosley, and methods
> of removing ink-stains from linen.[11]

Indeed for many women it may be the self-validation available through an
all-women group which is most important, a space where instead of being
'other', women are central and can explore issues, learn new skills, challenge
their own internalization of a socially constructed subservience and raise
both their confidence and their consciousness.

The regular meeting of a large group of women in a village in the 1920s
or 1930s in itself would necessitate a change of perceptions about male and
female spheres. This was magnified when the women, once a month, took
over the local pub (as they did in Singleton, West Sussex), a fairly significant
appropriation of male space. In other areas Institutes were instrumental in
building a WI or a village hall which became a female-controlled social
centre for the village, an alternative to the male-dominated pub. A village

hall built by the WI was able to become a physical manifestation of women's subculture.

The WI also ran a series of social events for the whole village. Their socials, held at Christmas time and sometimes at other times of the year, were very popular. They included entertainment by members, supper, and dancing. Yet there is every indication in humorous articles written in *Home and Country* that the men who attended these events (only if accompanying WI members) found them uncomfortable. They found the gender relations implied in women-controlled events difficult to handle. Within the Institute social, men were marginalized, they became outsiders, visitors within a female-defined culture and value system. An article entitled 'That Social' in *Home and Country* written from a male point of view and in their own particular form of rural dialect began:

> 'You could have gone an knocked me down with a feather' said Ephriam Pepperwot. 'When my ole Susan asked me to go to that there Women's Institute Social'.
> 'Thought you didn't allow husbands and such like heathen' I says.
> 'Not at our ordinary meeting we don't' says she. 'Still to the social husbands can come if accompanied by a member'.
> 'Not admitted except on a Leash' I says sarcastic.
> 'If you take it that way' says she.[12]

This article indicates that the female readership knew of the male discomfort and positively revelled in their sense of otherness.

The Women's Institute Movement and Domesticity

An organization like the NFWI may seem at a superficial level an instrument of bourgeois ideology trying to inculcate the working class with an image of womanhood that was domestic, patriotic, maternal, and dedicated to trying to live out some sort of rural idyll. This, however, would be to regard the membership as cultural dupes. The WI was deeply important to its members and as such deserves to be taken seriously, avoiding, in E.P. Thompson's words, 'the condescension of posterity'.[13] Many women never missed a meeting except literally for the birth of a baby or the death of their mother; others walked two or three miles to attend. Certainly the Jam and Jerusalem image makes the WI difficult to defend but also explains why they were able to be the 'acceptable' face of feminism. The organization did not challenge women's primarily domestic role but it certainly challenged its construction. There is a real problem with any sort of feminism which assumes that unless an organization rejects domesticity it can not be feminist. For the majority of working-class women the material reality of their lives was, and often still is, such that the world of work offers only women's work which was badly paid, low-status, part-time, and with limited promotion prospects. In the rural areas, where the WI operated, the problem of limited job opportunities was exaggerated by lack of transport. Work under these circumstances may have little to offer in terms of liberation – what it may offer instead is the double oppression of workplace and home. Domesticity and its evaluation are something which must be perceived as relative. Many of the women the WI was addressing itself to had passed the point of having any choice about whether to be unpaid domestic workers or not; consequently no movement which invalidated housewives' position would have gained much credence.

The WI perception of womanhood may have been primarily domestic but it was not a passive domesticity. It was a long way from the idealized 'angel in the house' popularized by Coventry Patmore in the nineteenth century.[14] A Lancing member writing in *Home and Country* in 1928 claimed: 'I hold no brief for the old fashioned, devoted wife, who scuttled about like a fussy hen, looking for little jobs and manufacturing them if they did not exist.'[15] Thus domestic labour is not seen as intrinsically beneficial and labour-saving devices are always valued by members. A Lancing member writing in *Home and Country* in 1930 praised the reductions in the clothes babies wore and in babies' feeds and new materials for clothes as all bringing significant and positive changes in housewives' lives. While this member saw domestic work as always primarily the woman's responsibility she was quite emphatic in her belief that domestic labour does not imply subservience, saying: 'Not masculine tyranny, not wifely submission, but this equal comradeship and the happy atmosphere are the real essentials of a lasting happy home'.[16]

The WI movement rejected the male capitalist value system's perceptions of their labour as of low status and value. Women were domestic workers and their work was equivalent to that of men. The movement did not merely validate women's work; it attempted to raise it to the level of skilled work, through competitions and exhibitions and by placing a monetary value on the products of female domestic labour when women sold items at WI markets and Institute sales tables. Skill is to a large degree a socially constructed concept and one that is primarily associated with men. Notions of skill are also linked to job satisfaction and ideas of fulfilment, something for which domestic labour justifiably has an ambiguous reputation. Acceptance or celebration of domesticity or domestic skills by the WI must never be taken as an acceptance of the gender power relationship or material circumstances within which they were predominantly carried out. Rather than being accepted, these were areas which the WI sought to change and renegotiate. For housewives, elevation of the importance of the home and domestic skills and an emphasis on the site of reproduction rather than production were both validating and empowering. They offered an alternative value system from which to challenge and renegotiate structures of male power.

It is from such an alternative set of meanings that some of the apparently traditional views on women's role should be seen. For example, in June 1928 Fernhurst WI in West Sussex had a debate on the subject 'Should husbands cook the Sunday Dinner?' It was decided 'they should not although wives would accept thankfully any help their husbands offered'.[17] At its most fundamental level this is women determinedly hanging on to one of the only power positions that they have, that of food production, and at that point in the week where it is given most status and centrality within family ritual. Alternatively the conclusion to the debate could be seen as a reflection of women's perception of their role as skilled and that like any other skilled person they were not going to diminish their skill by letting themselves be replaced by an unskilled labourer – even once a week. The mass appeal of the WI was based in many respects on their re-appropriation of domesticity and may therefore be seen as a precursor to recent Wages for Housework campaigns.

Being a skilled worker as opposed to a housewife was tied up with the right to time off work, i.e. leisure. This was an important role for the Institutes – a whole variety of areas developed: drama, craft work, folk singing and dancing, village socials, day trips and outings. These women-organized trips represented an appropriation of leisure previously unknown to rural women. Mrs Curly Brown of Somerset was heard to say: 'I have always wanted to travel and now it is going to be possible through the WI'.[18] However ideologically constructed the places for outings may have been (there were a lot of trips to stately homes), how they were perceived and the meanings placed on them remained firmly within the power of the members. When that same Mrs Curly Brown visited Cheddar Gorge in the 1920s it was not the stalagmites which impressed her but the electric lighting.

As a group of women who saw themselves as skilled domestic workers, members were concerned with improving the conditions of their labour and becoming more efficient workers. Autonomy and control over their lives were perceived as important, as were co-operative activities which would improve the material conditions of their lives. This led to involvement in a wide variety of social welfare campaigns, demanding analgesics for women in childbirth, district nurses, and improved rural housing.

Campaigns over Housing and Water

Welfare legislation is not, any more than anything else, open to a straightforward ideological reading. Its meanings, its uses, and the purposes it serves are different for different genders and social classes. Its introduction, implementation, and practice are the result of struggle along class and gender lines. The level of government intervention in the market economy is dependent on consensus support, and this consensus, or consent, is a site of constant struggle which like any other area of hegemony has to be constantly re-created or it can be lost, as recent years have indicated. In the struggles over building a consensus around the definition of 'adequate housing' the WI and other pressure groups played a significant role. The NFWI's motto was 'For Home and Country' and the ideological elevation of the home and pragmatic political action to improve the lot of rural women were intertwined in campaigns to improve rural housing, water supplies, and drainage. Their involvement in the contestation over 'adequate housing' indicates how the NFWI at both a local and national level was able to develop its own particular version of feminist politics in which the links between the personal and the political were always present.

Both water and housing were on the political agenda for the WI throughout the first forty-five years of their history; they were constantly the focus of national activity in terms of AGM resolutions, NFWI surveys and articles in *Home and Country*. At a local level Institutes undertook a variety of pressure-group-style activities. Warninglid WI in East Sussex had talks on subjects related to the home which included 'beauty in the home', whereas *Home and Country* included instructions on the best way to maintain an earth closet and demands in the 1940s for the nationalization of water and sewage.

The NFWI concern with housing operated at three levels. The ideological, that is issues around renegotiating the culturally held perceptions of the home and the domestic worker within it; the pragmatic, which was anything that housewives could individually do to improve the efficiency, atmosphere, or physical environment of the home; and finally political action to improve the material conditions of principally working-class homes. Some political demands, such as for builders to place water pipes inside or to lag them to prevent them freezing, were not applicable only to working-class homes.

Many social commentators, at both ends of the political spectrum, in the first half of the twentieth century, were keen to perceive the rural home as the embodiment of Englishness, the rural mother as the mother of the nation, and the NFWI therefore turned this cultural association on its head to use it to its own advantage, to demand improvements in social services and the material conditions of their homes. From the ideological significance placed by the WI on the home, their identification of domestic labour as both skilled and of high status, and their appropriation of notions of Englishness, they were able to claim an authority and validity from which to demand social reforms to improve rural houses. Sometimes their requirements were a little vague. For example, in 1921 an AGM resolution was passed which stated that 'The WI should support any legislation to elevate and purify the home life of the nation'.[19] Caroline Rowan has argued that the Women's Labour League and the Women's Co-operative Guild campaigns around housing were 'opening up the home as a legitimate arena of political struggle'[20] for women in the towns. The WI undoubtedly did the same for rural women. They were willing to join forces with other organizations in order to achieve improvements in housing, and the 1918 annual report noted that the NFWI co-operated with the Rural Housing and Sanitation Association and the Women's Labour League on housing campaigns.[21]

These activities could in a real sense operate to motivate politically women who might otherwise regard the political process as alienating. In a sense the personal may become truly political when women are involved in making political demands based on a desire to improve their own material circumstances in which they live and labour. The woman who had no piped water near her house, and therefore had to carry it all by hand, who sought to change this through combined action with other women, was surely similar to male trades unionists who attempted to improve the conditions of their labour. For these women, issues were raised about why the local authorities and central government would not prioritize their demands and what they did prioritize instead; this could be political consciousness-raising.

The 1919 Housing Act, while enabling local authorities to act and putting upon them a responsibility to do so, was open to very widely varying interpretation at local level. The implementation of the Act relied upon local council impetus, and this local Institutes tried to help along. In the immediate post-war period, Institutes were encouraged to gather information, for example on how many cottages were required by their village and why, the number of local people wanting cottages or imminently likely to, and the number of local cottages needing rebuilding or repair. Thus the WIs were taking on the role of building a consensus in the village around the need for council houses, as definition of need for housing was a fiercely contested area in the implementation of government housing policy.

In 1920 a writer in *Home and Country* pointed out that the Minister of Health had issued a circular expressing the hope that all local authorities would take steps to obtain the views of women with regard to the proposed housing schemes.[22] The article suggests that plans should be shown to women's organizations or that a special meeting of village women should be called and women co-opted onto housing committees. The intention was that after examination of the plans village women could send in their suggestions to the housing committee of the council. As the article in *Home and Country* argued, 'It is obvious that women who have to live and work and bring up their children in these houses know the requirements better than most men can possibly do'.[23] Institutes were encouraged to study the issues involved and appoint a special subcommittee to explore the issues of housing and transmit Institute views to local councils.

The demands of members were not by any means extravagant; they were encouraged by the leadership to work out and to express just what their requirements and priorities were. This very process encouraged them to perceive domestic labour and the material circumstances of their lives not as inevitable and fixed but as something constructed and not necessarily to their requirements. Such activities were essential in building up a consensus for government intervention and activity in the areas of housing and water supplies. They engendered in the members the possibility of change rather than an acceptance of the status quo. This could lead on to members questioning why things were as they were and what they could do about it. Consistent with this approach is an essay competition on 'The house I should like to live in' in 1919. It is important to note that the winner's requirements, which were a remarkable combination of the picturesque and the practical, were well below the recommendations of the government Tudor Walters committee. They were also dependent on the availability of piped water, which was a dream for many rural working-class housewives. Indeed, *Home and Country* ran an essay competition for members in the 1930s on the problems of lack of village water. Lack of piped water and sewage added tremendously to the burden of domestic labour which any rural housewife had to undertake and therefore it was an area of continual movement agitation. Demands for a water supply were seen as intrinsic to improved rural housing and in 1935 the NFWI tried unsuccessfully to get an amendment to the Public Health Bill to stipulate that no new plans for houses should be passed without a water supply.

A perception of the possibility of change is contained within all these Institute activities, and such a perception is an essential first step towards political activity. It can be argued that one of the reasons that women tend to feel alienated from the political process is because they do not see it as in any way likely to change their lives. This perception was something that the Women's Institute movement challenged. By making the home an area of political struggle the NFWI were able to politicize many rural women. The

WI's emphasis on the home, housing conditions, and water supplies was a natural extension of its motto 'For Home and Country' and reflected the predominantly domestic orientation of the membership. I have tried to show that this did not lead to a reactionary political position; rather it reworked ideas of the national importance of the home to empower women and demand social welfare reforms. For rural women in the WI, agitation around the home, as the site of their labour, was significant in raising their political consciousness.

Many of their social welfare demands, such as for water supplies to rural areas, in time came to be realized, but it is hard to estimate what level of direct responsibility they can take for this. Certainly the NFWI were significant in helping to build a consensus for the necessity of certain levels of rural housing and water supplies. Maybe the importance, for members, of campaigns for social reforms is more complex. They were an area through which members renegotiated the boundaries of their femininity and challenged their own internalization of the socially constructed subservience of women, and as such they could be a significant site of feminist activity.

Conclusion

With an organization like the NFWI as an example, perhaps a different perception of feminism in the past can be obtained. The women in the movement struggled in national political terms and locally to improve the material circumstances of women's lives. Perhaps more importantly, through the formation of a female subculture within rural villages, they provided a space for women to fight the internalization of male domination and to adopt an alternative value system, an example that feminism today would do well to remember. For my version of feminism they certainly provide an empowering and very suitable past.

Notes

1 P. Dudgeon (1989), *Village Voices*, London, Sidgwick and Jackson, p. 19.
2 *Ibid.*
3 E. Robins, *Votes for Women* (also published as *The Convert*, 1907), reprinted 1980, London, Virago.
4 R. Strachey (1928), *The Cause*, reprinted 1969, London, Virago.
5 *Home and Country*, October 1928, p. 524.
6 Northleigh WI History, in the NFWI archive.
7 NFWI Annual Reports, in NFWI archive.
8 Appuldram WI minute books, in West Sussex FWI archive.
9 Funtington WI minute books, held by Funtington WI.
10 *Home and Country*, May 1936, p. 281.
11 E.M. Delafield (1930), *Diary of a Provincial Lady*, reprinted 1984, London, Virago, p. 114.
12 *Home and Country*, January 1929, p. 32.
13 E.P. Thompson (1963), *The Making of the English Working Class*, London, Victor Gollancz, p. 12.

14 Coventry Patmore, 'Angel in the House' quoted in Carol Christ, 'Victorian masculinity and the Angel in the House' in M. Vicinus (Ed.), *The Widening Sphere* (1981), London.
15 *Home and Country*, June 1928, p. 212.
16 *Home and Country*, October 1930, p. 516.
17 Fernhurst WI minute books, in West Sussex FWI archives.
18 'Up-Along-Down-Along', members' memoirs of WI outings, in History file in NFWI archive.
19 National Federation of Women's Institutes (1981), *Keeping Ourselves Informed*, London, WI Books, p. 137.
20 C. Rowan (1982), 'Women in the Labour Party', *Feminist Review*, 12, p. 89.
21 NFWI Annual Report (1918), p. 9.
22 *Home and Country*, May 1920, p. 184.
23 *Ibid.*

Chapter 4

A Woman's Right to Work?
The Role of Women in the
Unemployed Movement Between
the Wars

Sue Bruley

Although unemployment was high for the whole of the inter-war years the periods of greatest distress were the years immediately after the collapse of the post-war boom, particularly 1920–1923, and those dominated by the depression of 1929–1934. Alongside these downturns the chronic decline of the old staples (coal, heavy engineering, shipbuilding, textiles) led to long-term misery for a great many of the workers associated with these industries and their families. The slump of the early 1920s hit very hard, creating desperate conditions among millions of families. The unemployed movement sprang out of this situation and was a powerful force which included several prominent women activists.[1] Regrettably we know little of the gender divisions within the unemployed movement in these early years. Reluctantly, therefore, this article concentrates on the period after 1923 for which much more evidence is available. My aim is to determine to what extent the unemployed movement responded to the specific needs of women, particularly married women, at a time when massive social pressures pushed them back into the home.

From 1921 the movement acquired some sort of national coordination under the title of the National Unemployed Workers' Committee Movement (NUWCM, becoming NUWM in 1930) with a head office in London and full-time officers paid 30s a week. As the movement became more centralized the domination of the Communist Party (CP) became more evident. Although its orientation was towards practical objectives, the most important being 'Work or Full Maintenance at Trade Union Rates', it also embodied the principle that ultimately only a Communist solution could suffice.[2] Not surprisingly, prominent activists on the whole represented declining crafts and industries, for whom unemployment was a long-term problem. Wal Hannington, National Organizer, was a skilled toolmaker. Harry McShane, Scottish Organizer, was also an engineer. Sid Elias, who chaired the Executive, came from the South Wales steel industry.

Many of the leaders of NUWCM had their roots as activists in the pre-war shop stewards' movement, especially those who had been engineers. Despite being class-conscious and militant, typically they had very traditional views regarding the sexual division of labour and tended to view unemployment as a male problem. Wal Hannington's *Unemployed Struggles* is a deeply moving testimony but he only makes a few passing references to the plight of unemployed women.[3] Hannington's indifference to women workers is confirmed by oral evidence from women who came into contact with him: 'he wasn't exactly hostile, he just didn't think of women'; 'he wasn't the least bit interested in women'.[4] For the first and second National Hunger Marches Hannington refused requests from women to march.[5]

Against this patriarchal tendency we must pose a contrary one. As a Communist satellite organization the NUWCM was required to pay at least lip service to the notion of sexual equality.[6] Implicit in its programme was support for equality of unemployment benefit at a time when it was virtually universally accepted that it was right to pay men a higher rate. The movement also opposed compulsory retraining of women industrial workers for domestic service. Trying to breathe some life into these principles and organize unemployed women were a handful of NUWCM women activists,[7] the most prominent of these being Lily Webb in the early years and Maud Brown from the late 1920s.

Born in 1897, Lily Webb was raised in a large family of cotton workers in Ashton-under-Lyne. After seven years in the mill she moved to Macclesfield to become a bus conductress. She was soon caught up in the post-war slump and was propelled into the Communist Party and the unemployed movement. Dedicated to the women's side of NUWCM activities for a time, she was National Women's Organizer and travelled all round the country.[8] At the national conference in 1926 Webb and the formidable Fanny Deakin from Stoke-on-Trent successfully moved a resolution against compulsory domestic service training for women.[9] They called for other forms of industrial training for women and for grants to expectant mothers. Webb sent in regular reports on her activities to the Party press. Always a 'grass-roots' organizer, she never worked at the movement offices and appears to have led a rather exhausting, peripatetic existence with her husband, who was a Communist organizer. From 1927 Webb was abroad attending various international congresses and tours, but she resumed her unemployed movement activities in the early 1930s.

There are numerous surviving photographs of Maud Brown as she tended to be near Hannington on national demonstrations. She came from a large working-class family in North London and worked for the Post Office as a sorter. Brown was active in the Labour Party in Tottenham in the early 1920s and was moved by the plight of the unemployed. It appears that at some point during the 1920s she decided to give up her job and offer her services as a volunteer at the NUWCM offices. This decision was reached with the co-operation of her husband whose income could keep both of them. Realizing the weakness of the movement's activities with regard to women, Brown began to specialize in this area and sometime during the late 1920s (probably 1929) she became officially known as the Women's Organizer. Hannington, who worked closely with her for several years, had nothing to say about her in his numerous writings. There is, however, plenty of oral evidence to suggest

that she was a remarkable woman, combining the qualities of efficient organization with an ability to lead.[10] She had overall responsibility for organizing
the women's contingents of the National Hunger Marches, and she marched
with the women. She continued her NUWM work right through the 1930s.

So what did the NUWCM do to attract women? Lily Webb pressed for
equality of unemployment benefit with men but this was never a serious issue
in the 1920s.[11] The most pressing question of unemployed women was that of
enforced domestic service. Here they would get no joy from Labour politicians. Margaret Bondfield, as a Labour Minister, urged women to enrol for
domestic training, not just in order to become domestic servants, but, more
importantly, to become better housewives since homemaking was 'the one
great occupation for women in this country'.[12] In these circumstances it would
not be unreasonable to assume that unemployed women would look to the
NUWCM for leadership. This posed a problem of organization.

The Communist Party had a policy of establishing women's sections at
all levels of the Party apparatus, although in many areas this directive was
completely ignored.[13] Where women's sections did exist they tended to be
constituted by wives of Party members and not working women. Like the
NUWCM, the Party's industrial base was heavily weighted towards engineering, mining, and other male-dominated occupations, making a weakness with
regard to women inevitable. Within the unemployed movement there appears
to have been a change of line on the question of women in 1926. We know
from the activities of Lily Webb that certainly in the early 1920s some women's
sections were established. The 1926 conference, however, passed a resolution,
supported by Webb, against women's sections, arguing that women's questions
should be dealt with by the whole branch.[14] Possibly these NUWCM women
had witnessed the ghettoization of women's issues within the Communist
Party and sought to prevent this in the NUWCM, although to a large extent
this was already a reality. This policy was reversed again in 1929 when Mrs
Youle, from Sheffield, proposed that all branches with more than twenty
women should set up women's sections and that a women's headquarters be
established at the NUWM offices.[15] It seems very probable that Maud Brown
became Women's Organizer at this point and that she was assisted by a small
committee which met regularly.

With numbers of unemployed women rising fast the women's committee
was not short of things to do. District conferences of unemployed women
were organized and a decision was made to include a women's contingent on
the National Hunger March planned to start at the end of March 1930. Just
how much of a struggle it was to persuade the NUWM leadership to this new
departure is unfortunately a matter on which we have no direct evidence.
The women's group was small, numbering only twenty-two in all. One participant, Margaret McCarthy, later wrote that it was difficult to find volunteers for the march.[16] This is hardly surprising given that the women were to
be away from home for several weeks. McCarthy was young and childless
and it seems probable that many of the other marchers were in a similar
position. Mature women were perhaps also free of dependent children but
they would be more likely to find the physical demands of the march too
arduous. For women with young children the prospect of finding suitable
childcare would surely have deterred all but the most indomitable, who were

invariably CP activists. Besides McCarthy, who was a weaver from Accrington and a member of the Young Communist League (YCL), there were several other textile workers. Rose Smith, who was also present, was a rising star in the Party and had recently become a full-time organizer. Her background was that of a schoolteacher from Mansfield, Nottinghamshire. Smith left behind twin boys who were looked after by a friend. Mrs Youle, referred to previously, was also on the march. She was married to the Sheffield NUWM organizer and they had four children.

The small group of Lancashire women began with a demonstration in Burnley on 20 April and then went by rail to Bradford to meet the remainder of the contingent.[17] According to McCarthy, who sent reports to the *Daily Worker*, the women got a great send-off and 'huge crowds came out to see us on our way'.[18] With banners declaring, 'Under-Fed, Under-Clad, Under the Labour Government', they met with kind gestures on route in the form of refreshments and offers of help. In some areas they received support from the official labour movement, especially Labour and Co-op women. But in many places no such help was forthcoming. This is not surprising since the 1930 March was in the middle of the CP's ultra-sectarian 'Third Period' when they were calling Labour and trade union leaders 'social fascists' and the NUWM was widely recognized as a Communist-led organization. As a result accommodation was a major problem. Often resort to the workhouse was unavoidable. Like the male marchers they were instructed not to submit to the humiliating workhouse regulations for 'casuals'. At Sheffield, as arranged, they entered the workhouse after an evening demonstration. They were then unexpectedly required to be 'bathed and searched'. Fearing a humiliating ordeal if they stayed the women marched out.[19] Unable to rest because of possible vagrancy charges, the group tramped about in the rain until early morning when they finally stopped for breakfast in the Communist Party office.

As arranged, the women then proceeded by coach from Sheffield to Luton. At St Albans they were fortunate enough to be fed and accommodated by the Labour Party. Via Barnet and Holloway the women marchers made their way to link, as planned, with the male marchers for a big demonstration in Hyde Park on 1 May. As the women filed into the park the veteran socialist and feminist, Charlotte Despard, stepped forward and presented Maud Brown with a bunch of red tulips.[20] Since one of the aims of the march had been to embarrass the new (minority) Labour administration, the women stayed on in London for a few days in an attempt to contact prominent Labour women. Margaret Bondfield, now Minister of Labour, adamantly refused to see deputations from either the women or the men. A deputation of women did, however, get to see Jennie Lee MP and Women's Officer Marion Phillips, but they came away with nothing except sympathy.[21]

As trade conditions worsened the numbers of unemployed men and women rose rapidly. Conditions at the Labour Exchanges, which were not built for the large numbers now using them, were often deplorable.[22] In these circumstances many women flooded into the NUWM. The women's sections often held meetings outside the part of the Labour Exchange where women had to register. Parts of Yorkshire, particularly Bradford and Doncaster, London, and Fife seemed to press ahead most vigorously with women's activities, often with women's sections numbering over a hundred. The domestic service issue appears to have been the most dominant. This was emphasized further in November 1930, when Margaret Bondfield announced in the House of Commons that young women must accept 'suitable offers' of work in domestic service. This decision was met with indignation in the labour movement. Apart from the fact that service was a non-insurable occupation, union leaders in the depressed textile trades were shocked at the thought of their women and girls being forced to leave home for work in service.

Such was the climate at the time of the NUWM Conference in Bradford in February 1931. The necessity for extending NUWM activities to include women was an important theme. Maud Brown moved the successful motion on 'The Task Amongst Unemployed Women' which went further towards meetings the demands of unemployed women than anything previously discussed in the movement. Apart from decent facilities for women at the Labour Exchanges, it pressed for creches for women who were signing on or looking for work. Demands were made for equality of unemployment benefit and for equal pay for women workers. The need to build unity between unemployed women and women workers through assisting at women's strikes and other disputes was established. Also heard were demands for improved maternity and child welfare services and free school meals and clothing for the children of the unemployed.[23] It was hoped that men in the movement would assist in the building of the women's sections and that activists in the women's sections would serve on the main branch committees, thus avoiding any 'separatist' tendencies which might arise.

Within four months unemployed women were facing a new offensive. The government, needing to make drastic economies in unemployment benefit, decided to single out groups it saw as marginal members of the labour force. So besides married women, part-time and casual workers were included in the Anomalies legislation which came into force in the autumn of 1931. In August, the number of women and girls registered as unemployed stood at 673,724.[24] There were many thousands more on short time or not bothering to register. During the autumn thousands of married women were struck off the register or denied the right to register. *The Times* reported that within a few weeks of the Act becoming law Sheffield had disqualified 2,500 women, Tyneside 1,200 and Nottingham 800.[25] Inevitably, the textile towns where women traditionally continued to work after marriage were badly hit. In Bury it was reported that 2,000 women had been refused benefit in the six months following the Act.[26] NUWM activities were obviously affected by the legislation. The Kirkcaldy branch in Fife reported that its committee was 'very busy with Anomalies cases of married women'.[27] Many women were assisted by NUWM legal specialists when appealing against disqualification.

Although many women's sections were active during the height of the depression years, work among unemployed women was not always a great consideration for the branch as a whole. Where the majority of women were the wives of unemployed men the women's sections often took on a servicing role, organizing socials, decorating banners, and generally supporting the main branch. The women's committee at head office would periodically issue scathing criticisms to branches for not paying sufficient attention to unemployed women: 'no serious attention is being taken of the correspondence sent out by the women's department . . . the comrades fail to understand the importance of work among women'.[28] Whilst many NUWM branches dragged their feet, occasionally there were cases of outright opposition. In Plymouth it was reported that despite having over 200 women members the branch had refused permission to set up a women's section. A deputation to see the Mayor did not include women 'until they forced themselves into it'.[29]

The build-up to the National Hunger March of 1932 took place against this background. Apart from the Anomalies legislation, the impetus for the march came from the new 'means and needs' test and the across-the-board 10 per cent cut in benefit. The women's contingent set off from Burnley on 9 October, with the aim of a hundred women marchers not achieved by a long way. Many of the forty women (two of whom did not make it to London) were cotton weavers. The cotton industry was in deep crisis, with its workers, pre-eminently women, fighting the 'more looms' system, which created unemployment, and wage reductions. Weavers, paid on piece rates, who could find work were often forced to work on less than the usual four looms for wages which were spiralling downwards. This left them no better off than on the dole.[30] Maggie Nelson, a weaver from Blackburn, inhabited this milieu in which work and unemployment were constantly overlapping. A deeply committed communist, she participated in the national marches of 1934 and 1936 as well as that of 1932. Marion Henery, one of the Scottish marchers, who was a YCL representative, remembered Nelson as a 'very strong personality' who was very successful at addressing street meetings. She 'made an appeal to them, won their hearts and got great collections'.[31] Nelson was postwoman to the women's contingent and photographs show her distributing letters to the other women. She was separated from her husband and left her three children with friends whilst on the marches.

The women marchers ranged in age from 16 to 62. They marched in step, always led off by a Mrs Oswaldtwistle, with the clogs of the Lancashire women clanging on the roads. Under the banner 'We Want Bread' they covered all of the 250 miles to London. Lily Webb, who was joint leader with Maud Brown, shouted slogans against the means test and the Anomalies Act.[32] Maggie Nelson remembered chanting

Work Work Work
We want work
And an end to the means test
Slave camps and the rest.[33]

Although carrying packs of forty pounds and often tramping in terrible conditions the marchers kept up their morale. Trying to survive on thirteen shillings and sixpence a week had been very grim so many ate better on the march than they did at home, especially where reception committees rather than the workhouses had laid on meals. They sang whilst marching to help pass the time, the Internationale being a favourite, but also sang popular tunes such as 'Who were you with last night?'

Although the Communist Party was still in its ultra-left 'Third Period' the women appear to have received quite a lot of help from labour organizations. Lily Webb made a point of mentioning this.

Reception committees awaited us in the towns we stayed for the night and often they marched out to meet us and carry our packs. Sometimes they came with bands and banners and always a great crowd awaited us. . . . The Reception Committees organised by the NUWM were many times joined by other working class organisations, including the Co-operative Women's Guilds, whose numbers played a big part in a number of the reception centres, especially in the collection of funds, etc. as did women members of the CP, ILP, TUs etc.[34]

Nevertheless it was still necessary to make use of the facilities of the workhouse for some of their accommodation. Again the women demanded to enter on their own terms. Apparently this was accepted in Rotherham and Derby without too much difficulty, but at Burton-on-Trent the marchers were told to report inside by 8 p.m. The women ignored this instruction and held a meeting in the market square until 9.30. Large crowds then followed the women to the workhouse where admission was refused until they agreed to submit their names. Fearing victimization the women refused and continued to agitate outside until nearly midnight when they were finally admitted with all conditions removed.[35] Fortunately the rest of the march proceeded without serious incident. At St Albans, as in 1930, the marchers received generous hospitality, this time under the auspices of a reception committee set up by the local trades council. The committee organized evening entertainment, accommodation, meals, boot repairs, spare clothing, and baths.[36] After eighteen days on the road the marchers eventually arrived in London in time for a mass rally in Hyde Park. Lily Webb spoke on the leading platform with Wal Hannington and Harry Pollitt. The rally ended acrimoniously with a police charge resulting in seventy-seven casualties and numerous arrests, including Webb.[37]

The fourth national march achieved much publicity for unemployed women and many of the marchers continued their activities after February 1932 in their own localities. A number of local marches, some of which involved women, were organized during 1933. The Scottish march to Edinburgh in June included a women's section. Maud Brown sent in reports on unemployed women in the *Daily Worker*. Periodically, she would remind readers of the numbers of married women who had been struck off the unemployment register. In December 1933 she reported that the total had reached 203,427 and we know this to be substantially correct.[38] The NUWM published several pamphlets in the early 1930s geared towards helping claimants to understand the new unemployment benefit legislation. The Anomalies legislation was regarded as particularly pernicious as it denied married women benefits which they had paid for in contributions. *The Practical Guide to the Unemployment Acts* and its revised versions, which were mainly written by Sid Elias, advised married women on the best way to approach officers at the Unemployment Assistance Board (UAB) office.[39]

On a grass-roots level, however, things were not looking so good. There were complaints about the lack of women speakers. In response to this the Women's Department began to issue Speaker's Notes which were full of useful data and arguments. Many women's sections had failed to thrive and those that did function did not do so in the way that was originally intended. One indication of this is the question of social centres, where there was a clear need for the NUWM to maintain morale and keep members away from the government-sponsored social centres. A circular urged branches to organize whist drives, concerts, and sports competitions. Although it was stressed that this work 'must not be relegated simply to the women's committee' it is implied that this was a common occurrence.[40] Finally, the period between the fourth and fifth national hunger marches was characterized by the emergence of another issue. In November, 1933, the British Medical Association published a minimum diet for health. Even though the diet was largely based on cereals and potatoes, it did not take the NUWM long to work out that it was beyond the reach of most unemployed families.[41] This new evidence of the link between unemployment and malnutrition was combined with reports of rickets among children of the unemployed in Glasgow and it was quickly taken up by the Women's Department. Ultimately it was to transform the whole nature of their work.

From the autumn of 1933 plans were made for another national march linked to a Congress of Action in London. The main demands were abolition of the means test, restoration of the 1931 cuts, a seven-hour day, and work at trade union rates. The most pressing question, however, was the need to provide organized opposition to the Unemployment Bill which, amongst other things, rendered permanent the 1931 cuts in benefit. The women's contingent, which numbered forty-eight, set off from Derby on 12 February 1934. Mrs Chater, from Gateshead, whose husband had been unemployed for four years, led the women. Maud Brown and Maggie Nelson were amongst the other women marchers, who represented all the major depressed areas.

The women, many wearing distinctive red berets and dresses, marched via Coventry, Northampton, Bedford, and Barnet to London for a mass rally in Hyde Park where Hannington reports that a 'special cheer' was given out when the women made their entrance.[42] Besides the main women's contingent there was a small group of women marching with the South Wales contingent, who were greatly encouraged by the march leader and South Wales NUWM organizer, Lewis Jones. The group of about fifteen women marched proudly at the head of the march and were led by Dora Cox, wife of the Party organizer in south Wales, Idris Cox. Cox recalled that several of the women marchers were in middle age.[43] There were a number of young women in the group. Apparently, none of the Welsh women marchers had small children. Not surprisingly, since there was very little work for women in the Rhondda, many of the women were wives of unemployed miners. According to Cox, there were excellent reception committees awaiting their arrival in Newport, Bristol, Bath, Swindon, and Slough. The local Hunger March Solidarity Committee would prepare food and accommodation and advertise ('chalk up') street meetings. At some of these meetings, for example that at Slough, the women marchers who spoke had never spoken at a public meeting before.

The Communist Party was by this time moving away from its ultra-left policies and calling for united front activities. The 1934 march certainly attracted much more labour movement support than previous marches. Many Labour and Co-operative women provided meals for the marchers en route.

Labour movement support was also evident in the women's contingent which attracted a gift, amongst others, of over £7 from the Association of Women Clerks.[44] A deputation of women, led by Maud Brown, went to see Ishbel MacDonald, daughter of the Labour leader. The women, all but one of whom had been working until becoming unemployed, were appalled by Ishbel MacDonald's remark that the training centres were 'quite good' and responded with 'why don't you go into domestic service and see how you like it.'[45]

The Congress of Action, which was held in Bermondsey Town Hall, included many women speakers and reiterated many of the demands made in 1931, including creches and 'equal pay for equal work'. Among the various publications associated with the Congress were two pamphlets specifically aimed at women, *The Unemployment Bill and Women in Industry* and *Women and the Slave Bill*.[46] The women stayed on in London for ten days and attended numerous rallies and public meetings. Both Mrs Chater from Tyneside and Mrs Brown from South Wales proved to be very effective speakers. In the budget of April 1934, came the news that the 10 per cent cuts, imposed in the economy measures of 1931, would be removed.

During 1934 the women's sections of the NUWM appear to have been in a fairly static position. The Women's Department reported in April that 'our women's sections are still weak and in many places no efforts are made by our branches to organize these sections'.[47]

Towards the end of the 1934, however, there was an upsurge in activity which continued into 1935. The primary cause was the proposed implementation of Part II of the Unemployment Act, which abolished local Public Assistance Committees and put unemployment relief on a national rather than local footing. Unfortunately the proposed new relief scales meant a two-shillings-a-week cut for a married couple and drastic cuts of three to nine shillings for other adult men and women still living in the parental home. On 4 February a mass demonstration of women in Merthyr was held outside the local UAB office. According to Hannington,

> Whilst they were outside the offices of the UAB somebody pointed to an official who was looking out of the window and who appeared to be sneering at them. This incensed the women ... instantly a rush took place leading to the smashing of the windows and doors of the UAB office. The police made several arrests, but the women fought with such determination that they compelled the police to release every one of those under arrest. Only with the greatest difficulty were the police able to protect the UAB officials from violent assault.[48]

The following day further demonstrations were held in Scotland, Lancashire, Nottinghamshire, Derbyshire, Cumberland, Cheshire, and the Midlands. At the same time the government announced that the new scales would be withdrawn and reductions in benefit already imposed would be repaid. On 8 March thousands of Rhondda women marched to the local UAB offices demanding abolition of the means test and two months later 300 Rhondda women marched to meet the Prince of Wales who was visiting Cardiff. They were only stopped at Pontypridd after they had broken through three police cordons.[49]

During this period the number of NUWM women's sections grew rapidly with seventeen new ones being formed between February and May 1935.[50] The issues which dominated these women were, apart from relief scales generally, malnutrition, maternal and infant mortality, and related campaigns for milk and school meals. During 1935 the NUWM published a pamphlet by Maud Brown, *Stop This Starvation of Mother and Child*, which drew attention to the very high levels of maternal mortality, maternal morbidity and infant mortality in the depressed areas.[51] These issues were being raised by other bodies besides the NUWM such as the Committee Against Malnutrition and the Children's Minimum Committee.[52] Besides these national pressure groups there were many local groups which attracted broadly based support, often with NUWM participation. In this age of Popular Fronts the CP and the NUWM stressed the need for 'all progressive forces' to rally together against the evils of malnutrition and its deadly consequences. Against such formidable lobbying, and because of the gradually improving economic position, the government quickly began to make concessions.

But what of the unemployed woman? The advice service to the unemployed continued as before, including advice to women who had been disallowed under the Anomalies Act or who were refused benefit for refusing training for domestic service. But, on the whole, it was clear by 1935 that the main thrust of the NUWM's work with women was geared towards the wives of unemployed men. The NUWM was in any case in decline as economic growth began to reduce the national figures for those out of work, although the depressed areas continued to suffer. The Communist Party was growing but its chief concerns by 1936 were Spain and the threat of fascism. On the industrial side it was slowly dawning on the Party that many of the 'new industries' around London and in the Midlands used young, female labour with no traditional loyalty to either trade unionism or the Communist Party.[53]

The 1936 march, now called the 'National Protest March', was something of a reminder to a nation whose living standards and expectations were rising that not everyone was sharing in the new prosperity. It is an indication of the decline of the movement that the women's contingent dropped to thirty-two in 1936 from the 1934 height of forty-eight. The women assembled at Coventry and proceeded towards London for a rally on 8 November. As usual Maud Brown and Maggie Nelson are recognizable amongst the march photographs which have survived. Also present was Bertha Jones, with her drum, which the women sang to as they marched. One of their chants went

March along, working women, march along,
In the ranks of the workers you belong.[54]

The call for an end to the hated means test and for work at retraining centres at trade union rates won the marchers widespread support. Mayoral welcomes were not uncommon, some occurring in towns controlled by the Conservatives. Despite the fact that some of the women marchers were themselves unemployed the theme for women was more than ever geared towards the wives of unemployed men. As the article in the CP women's paper stated, 'the women are marching in defence of their homes and families'.[55] Whilst the marchers were still in London a large public meeting was held to launch the pamphlet, *Unemployment and the Housewife*, which was issued by the Committee Against Malnutrition but widely circulated in the NUWM. Prominent women such as Edith Summerskill joined in the protest on behalf of the women who had to keep a family together on unemployment benefit. Maud Brown argued for a Housewives' Minimum and urged women to press for claims under a 'pots and pans' clause in the Unemployment Act.

After the 1936 march there were no more national marches and the NUWM gradually faded. Conferences continued to issue rhetoric about unemployed women but in reality they were a very minor concern. At the last conference before the war, and the end of the NUWM, 140 delegates were present, only seven of whom were women.[56]

Conclusion

Given the large numbers of women who became unemployed, particularly from the textile trades after 1929, why did women not become more central to the NUWM and why did it not recruit more unemployed women? Nominally the NUWM had a policy of sexual equality, but in practice it was imbued with an aggressively patriarchal ideology and culture. Only Maud Brown and Lily Webb appear to have made a serious attempt to check this male domination of the leadership. During the period around the time of the Anomalies Act the NUWM did respond to the large numbers of women becoming unemployed. But the introduction of women's sections was clearly not successful from the point of view of women workers. In areas such as Lancashire women workers did not join such sections enthusiastically. As in the Communist Party, the women's sections were not successful in promoting women workers but they did succeed in areas where they overwhelmingly consisted of the wives of unemployed men.[57] The largest NUWM women's section was in Sheffield with 'nearly 300 members', many of whom were the wives of engineers.[58] The most militant women were the wives of unemployed South Wales miners. The NUWM reinforced the prevailing sexual division of labour and failed to promote the interests of women workers. Perhaps this was inevitable given the nature of the trades which founded the NUWM in the early 1920s and continued to dominate it in the period when many women became unemployed. At least they made some kind of effort, though, which is more than can be said for most of the labour movement, most notably the Labour Party, in these years.

Notes

I am grateful to Dora Cox and Peter Kingsford for help with this article. The opinions expressed are entirely my own. Thanks to Lorraine Pare for her speedy and efficient typing. Thanks also to Barry Edwards for his patience.

1 Lilian Thring, who edited the London paper *Out of Work*, took a prominent role in the occupation of Essex Road Library in North London. For details see the entry on Thring by K. Weller in J. Saville and J. Bellamy (Eds) (1987), *Dictionary of Labour Biography*, Basingstoke, Macmillan, vol. 8. Mary Bamber was also a

well-known figure and leading activist in the storming of the Liverpool Art Gallery in 1921. Some information on Bamber can be found in Jack and Bessie Braddock's joint autobiography, *The Braddocks* (London, Macdonald, 1963). Bessie Braddock (née Bamber) was the daughter of Mary Bamber.

2 The programme of the first National Hunger March in 1922 included a statement about long-term aims which read 'the general uplifting of a working class as a whole until emancipation from the clutches of a decaying system of society based upon Rent, Interest and Profit is achieved'. P. Kingsford (1982), *The Hunger Marchers in Britain*, London, Lawrence and Wishart, p. 34.

3 For example, *The Problem of the Distressed Areas* (London, Gollancz, 1937) has a chapter on retraining centres ('Slave Camps') which devotes one paragraph to the problems for women undergoing demestic service retraining. In *Unemployed Struggles* (London, Lawrence and Wishart, 1936), p. 216, Hannington refers to the Anomalies Act but fails to draw attention to its great significance for married women or to elaborate on its consequences.

4 Interview with Katie Loeber (previously Kant), 23 April 1978, London. When Mrs Loeber was married to Ernest Kant he had a period working for the NUWM and as a result she got to know Hannington quite well. Interview with 'X', 9 August 1977, London. 'X', who does not wish to be identified, worked as a clerical assistant in the NUWM offices.

5 I. MacDougall (1990), *Voices from the Hunger Marches*, Edinburgh, Polygon, vol. 1, p. 4. This new collection of recollections of Scottish marches is welcomed. I have included material on Marion Henery, the only woman in vol. 1, in this article. I understand that vol. 2 will have more recollections from Scottish women marchers.

6 For details on the Communist Party and women see S. Bruley (1980), *Socialism and Feminism in the Communist Party of Great Britain, 1920–1939*, PhD thesis, University of London; republished (1986) as *Leninism, Stalinism and the Women's Movement in Britain, 1920–1939*, Garland Press, New York.

7 This article concentrates on women's issues in the unemployed movement. Consequently, there is little material on the women who were active within the NUWM but who did not seem especially interested in the particular problems of unemployed women. An example is Kath Duncan, who was a very well-known figure in Deptford, South-East London, during the 1930s.

8 Lily Ferguson (Webb), *Some Party History*, n.d. (1960s?). I obtained a copy of this autobiographical sketch from James Klugmann. I presume that the Communist Party archive holds the original.

9 *Report of the 5th National Conference*, NUWCM, 23–25 Jan. 1926, Stoke-on-Trent, pp. 23–4.

10 Interviews with Katie Loeber (see note 4); Dora Cox, 10 May 1977, London; Hilda Vernon, 13 March 1977, London. All the NUWM material quoted here is in the Marx Memorial Library, London.

11 See, for example, an article by Lily Webb for *Working Woman*, October 1927, in which she argues for an allowance of thirty shillings for men and women. But, overall, the references to this issue are scarce.

12 M.A. Hamilton (1924), *Margaret Bondfield*, London, Leonard Parsons, p. 171. Later, in her second period of office, Bondfield was fond of opening retraining centres for domestic service.

13 S. Bruley, *Socialism and Feminism in the CPGB*, especially ch. 3.

14 *Report*, 1926 (see note 9), p. 24.

15 *Report of the 6th National Conference*, NUWCM, 19–20 October 1929, p. 22.

16 M. McCarthy (1953), *Generation in Revolt*, London, Heinemann, p. 152.

17 Kingsford (1982), p. 124, writes of the women setting off from Barnsley, when he must mean Burnley.

18 McCarthy (1953), p. 152.
19 McCarthy (1953), p. 153.
20 *Daily Worker*, 3 May 1930.
21 Kingsford (1982), p. 126.
22 Louie Davies described what it was like to 'sign on' every day in Bolton. 'You'ould go out at morning at half past eight to be at dole for nine ... You'ould be all the way stood up the steps, all the room would be a solid mass of people. There'd be four or five hundred in this room' (Interview, 5 September 1977, Lancashire).
23 *Report of the 7th National Conference*, NUWM, 21–23 February 1931, Bradford, pp. 16–17. For evidence of a similar approach in the Communist Party see S. Bruley, *Socialism and Feminism in the CPGB*, and S. Bruley, 'Gender, Class and Party, The Communist Party and the Crisis in the Cotton Industry between the Two World Wars', *Women's History Review*, vol. 2, no. 1, 1993.
24 *The Times*, 26 August 1931.
25 For these figures and general comment see *The Times*, 14 November 1931; 16 November 1931; 25 November 1931.
26 *Cotton Factory Times*, 11 March 1932.
27 *Bulletins on Branch Activity*, NUWM, monthly, October/November 1931.
28 National Administrative Council (NAC), NUWM, 7 and 8 May 1932, Women's Department Report.
29 NAC, NUWM, 3–4 October 1931.
30 See S. Bruley, 'Gender, Class and Party', for more about women weavers.
31 I. MacDougall (1990), p. 51. The Lancashire woman that Marion Henery is referring to, who came from Nelson or was Mrs Nelson, is without doubt Maggie Nelson.
32 Kingsford (1982), p. 143.
33 This featured in an article on Maggie Nelson in *Socialist Worker*, 18 Dec. 1976, and subsequent Interview, 11 March 1977.
34 Webb, *Some Party History*, p. 143.
35 This incident is reported in the *Derby Evening Telegraph*, 18 Oct. 1932, and is referred to in Hannington, *Unemployed Struggles*, pp. 244–5, and Kingsford (1982), p. 146.
36 Kingsford (1982), pp. 143–4.
37 *Daily Worker*, 1 Nov. 1932; *The Times* 2 Nov. 1932.
38 *Daily Worker*, 3 Dec. 1933. A. Deacon (1976), *In Search of the Scrounger*, London, Bell, p. 82, reports that by April 1933 over 200,000 claims by married women had been disallowed by the new regulations. In the same period 45,000 seasonal workers, 5,000 intermittent workers, and no part-time workers were disallowed, which clearly reveals that the main thrust of the Act was directed at married women.
39 NUWM (n.d.), *A Practical Guide to the Unemployment Acts*.
40 *Memo on Social Life in the NUWM* (n.d.) (After Easter 1933).
41 N. Branson and M. Heinemann (1971), *Britain in the Nineteen Thirties*, London, Weidenfeld and Nicolson, has a brief account of this controversy, p. 229.
42 Hannington, *Unemployed Struggles*, p. 287.
43 The information in this paragraph was obtained from the interview with Dora Cox (see note 10), and from a tape of her speech at the Lewis Jones Day School, organized by *Llafurr*, 25 Nov. 1978, in the Rhondda Leisure Centre.
44 Kingsford (1982), p. 185.
45 *Daily Worker*, 10 March 1934.
46 Also circulated was a pamphlet written by G. Halkett, and NUWM member from Dumbartonshire, *Why No Work Schemes for Women?* Reported in the *Daily Worker*, 23 Feb. 1934. Does a copy survive?
47 NAC, NUWM, Women's Department Report, 7–8 April 1934.

48 Hannington, *Unemployed Struggles*, p. 312.
49 Speakers Notes, Women's Department, NUWM, 14 May 1935.
50 NAC, NUWM, Women's Department Report, 25–26 May 1935.
51 For an account of these problems see J. Lewis (1980), *The Politics of Motherhood*, London, Croom Helm.
52 The Committee Against Malnutrition, which consisted of pro-CP medics and other health experts, issued its first bulletin in March 1934 and pronounced its concern about the widespread undernourishment of families of the unemployed and low-paid. Bulletins of the Committee Against Malnutrition are housed in the Marx Memorial Library. The Children's Minimum Council, which emanated from the feminist movement, most notably Eleanor Rathbone MP, pressed for free milk, school meals, and increased benefit allowances.
53 There is an interesting article by Rose Smith, 'Women in Industry', *Daily Worker*, 31 Jan. 1936, on this question. For a recent account see M. Glucksmann (1990), *Women Assemble*, London, Routledge, and M. Glucksmann, 'In a Class of Their Own? Women Workers in the New Industries in Inter-War Britain', *Feminist Review*, 24, Autumn 1986.
54 *Daily Worker*, 7 Nov. 1936.
55 *Woman Today*, Nov. 1936, p. 11.
56 *Reports and Resolutions Passed at the 11th National Conference*, NUWM, Blackpool, 28–29 January 1939, inside front cover.
57 For details on CP women's sections see S. Bruley, *Socialism and Feminism in the CPGB*.
58 Circular Number D30, NUWM, 4 Jan. 1935, is a report on the progress of women's sections. It quotes Sheffield as 'perhaps the largest section' with 'nearly 300 members' (p. 1). There were a total of twenty-eight functioning women's sections at this date.

Chapter 5

The Culture of Femininity in Women's Teacher Training Colleges 1914–1945

Elizabeth Edwards

This chapter concentrates on three teacher training colleges – Homerton, Bishop Otter, and Avery Hill. I want to discuss how feminine ideology with its hegemonic domestic and familial values continued to form throughout our period the enveloping context of the work of these colleges. The two voluntary colleges – Homerton and Bishop Otter – had been training women teachers since the second half of the nineteenth century. Avery Hill, on the other hand, was a new maintained college founded by the London County Council in 1906.

Homerton's flagship status in the teacher training world was due both to the outstanding leadership of its first two women principals and to its location in the university town of Cambridge. Mary Allan, who was principal from 1903 to 1935, began immediately to raise the academic and social status of Homerton by appointing women graduates to her staff, and by personally selecting students who could benefit from the enhanced academic teaching and cultural socialization which a graduate professional staff could provide. Cambridge's exceptionally good cultural and recreational facilities were an added attraction to both staff and students. Moreover the opportunity to meet eligible future husbands from the university élite – especially towards the end of the period when relations between the university and Homerton had become less distant – provided an important hidden curriculum for both students and their parents (Edwards, 1992).

Bishop Otter on the other hand was always hampered by its location in the small cathedral city of Chichester, where 'more than elsewhere perhaps student life is divorced from the movements of the larger world outside the little Cathedral City' (*Bishop Otter Magazine*, 1922). Moreover as a Church of England college it was under the aegis of clerical and patriarchal values. It was only for instance with reluctance, and under 'strong pressure' from the Board of Education, that the governors in 1920 appointed the college's first woman principal (*Bishop Otter Magazine*, 1971). Avery Hill was a much larger college than either Homerton or Bishop Otter with a student population of 300 to their 200, but not all its students were resident. The college always enjoyed a progressive regime which owed much to its enlightened government by the London County Council and to students' access to the metropolitan culture of London (Shorney, 1989).

Up to 1945 only a small minority of girls proceeded beyond the elementary school to receive a secondary education, and of those that did an even smaller number then went on to college (Beddoe, 1989). However for those girls who did continue their education beyond the secondary stage, a teacher training college was by far the most popular destination. Women teachers had predominated in the elementary schools since the beginning of the century. In 1914 they formed 75 per cent of the elementary teaching force and in 1938 the proportion stood at 71 per cent. There are interesting gender and class implications in the popularity of elementary teaching as a career for girls. Teaching was seen as an extension of the service orientation of the domestic and familial values of feminine ideology. Moreover, unlike other careers which became available – in theory at least – to girls after World War One (e.g. the legal profession and the higher civil service) elementary teaching as a predominantly female preserve did not oblige girls to challenge or compete with men. And yet, as Summerfield (1987) has shown, the academic staff in girls' secondary schools, most of whom were graduates, considered the two-year teacher training course definitely inferior to three years at a university:

We always used to feel that [the teachers] were biased towards university and latterly as you went back, they really only seemed to be particularly interested in university-trained people. Perhaps that's because they were on common ground. They were all graduates. (Summerfield, 1987)

The academic staff in training colleges were also graduates, and it was their transmission to students of the values and traditions of their own university education which was a vital element in the cultural enrichment which the college experience offered women.

Widdowson has shown that in the nineteenth century elementary teachers were in the main clever working-class girls. After their own elementary education was complete at the age of 14, girls intending to teach served a four-year apprenticeship as pupil teachers before qualifying for free tuition at a training college. Pupil teachers received some post-elementary education at pupil teacher centres but they had no secondary schooling as such. By the first decade of the twentieth century, however, the increase in the number of schools for girls greatly widened the opportunities for secondary education. This allowed the pupil teacher system to be gradually phased out to be replaced by a student teacher year between the completion of secondary education and entry into a training college. The improved standard of academic qualifications for entry to training college was crucial in making the teaching profession more attractive, particularly to clever and ambitious girls from the lower middle class (Widdowson, 1983). The first students at Avery Hill in the early years of the century 'came from the Edwardian petite bourgeoisie. They were the daughters of tradesmen, artisans and small employers, of public employees, including teachers, of clerks and small shop-keepers' (Shorney, 1989).

Beddoe has pointed out that the class origins of girls going to teacher training college between the wars remains unclear (Beddoe, 1989). The evidence from Homerton however would seem to indicate that, throughout the period, students continued to be predominantly clever girls from lower-middle-class homes who had received secondary education at maintained grammar schools. Of the ninety-six students entering Homerton in 1914, 88 per cent had had teaching experience which qualified them for free tuition at college. Seventeen per cent had been pupil teachers and 71 per cent student teachers. Ten years later the proportion of entrants with previous teaching experience had only dropped to 80 per cent but by the end of the 1920s there had been a dramatic change. In 1929 only 32 per cent of Homerton entrants had had teaching experience (5 per cent as pupil teachers and 27 per cent as student teachers) and the number of girls who had received a full secondary education culminating in the award of the Higher School Certificate had doubled from 19 per cent in 1924 to 38 per cent. This move towards a full sixth-form education for college entrants had already been recognized and approved by Bishop Otter's principal:

> Higher Certificate applicants are on the increase. Some of the best LEAs are discontinuing or discouraging the Student Teacher year so students are coming straight from the 6th form with higher standards. They compare well with those who have been student teachers in their college results. (Bishop Otter Annual report, 1925–1926)

By 1939 the proportion of student teachers at Homerton had dropped to 13 per cent and those with Higher School Certificate had risen to 62 per cent. Interestingly, though, 20 per cent of entrants had not completed a full sixth-form course and entered college with only the School Certificate. This helped to reinforce the continuing second-class status of training colleges (Homerton College Registers, 1914, 1924, 1929, and 1939). By this time a few 'sophisticates and glamour girls' (HCA, Acc. no. 1403, no. 3, 1935–1937) were entering Homerton from private schools or from overseas – but the overwhelming majority remained lower-middle-class girls who had been educated in the public sector. As one student remarked, 'We were nearly all hard up. Only the Bermuda girls had banking accounts and beautiful clothes' (*Homerton Roll News*, 1988–89). A member of staff in the early 1940s confirmed that 'it was rare for public school girls to come to Homerton' (HCA, Acc. no. 1412).

I now want to examine in closer detail three aspects of the culture of femininity in teacher training colleges: the translation to an institutional setting of the familial and domestic customs of the middle-class home; the widening cultural horizons which the experience of college afforded girls; and lastly the issue of sexuality which by the end of the period had become an increasingly contested arena.

Familial Customs

The organization of students into 'mother and daughter' pairs was an effective method of helping girls, many of whom had never been away from home before, to settle into the new environment of college life. Each first-year student was allotted her own mother from among the second-year students.

Mothers supervised their daughters' socialization into college life and acted generally as their mentors and friends. Similarly, when daughters entered their second year they in turn became mothers to the next generation of students (Edwards, 1990). Students' own words reveal how effectively this replication of family relationships eased the passage from home to college:

> The family system goes a great way here. To feel that there is one in the college to whom you belong, one who is to be your 'mother' as a helper and guide makes you feel that Homerton is just a huge home and you yourself are one of the big family. (*Homertonian*, 1915)

> There to welcome me was my college mother, who showed me to my room and then we made a tour of the building, learning about regulations as we moved around. It was an adventure for me because it was the first time of leaving Home. The next day she invited me to accompany her to Church. (HCA, Acc. no. 807)

Another Homerton student, writing home to her family in 1928, revealed not only how important the mother/daughter discourse was to students' initial socialization into college life, but also, that, in time, as in the home, daughters would move away from relying solely on their mothers to form their own relationships:

> Three girls opposite to me are very unhappy but they haven't got a nice college mother. [Mine] is awfully nice and we are always together

and later:

> I'm so busy that I don't go with her [her mother] except to church but I see her at mealtimes. I don't need her care now. I've got [two other girls with whom she remained life-long friends]. (HCA, Acc. nos. 889–895, 1928)

At Bishop Otter family relationships and the mother/daughter discourse in particular at first proved more problematic, reflecting the college's stricter and more old-fashioned ethos. One student described her arrival at college in 1920 and her welcome by the principal as follows:

> My heart continued to thump and bump most uncomfortably. We were marshalled into a single line and a voice hissed into my ear 'Walk up the steps, curtsey and say your name'. The line moved forward, now was the moment. 'Miss. . . ., Ma'm' I stuttered. 'I welcome you into our family.' (BOA, G78, 1920)

During the 1920s, however the, 'slavery' of the Bishop Otter system which 'replicated only too well the power relationship of the mother/ daughter relationship in the home' was gradually modified until it matched the Homerton arrangements where 'each second year takes a newcomer under her wing . . . but the adoption lasts for a short time as long as the two parties concerned like to make it'. The old system had encouraged an over-exclusive interpretation of the mother/daughter discourse which was inimical to the cultural freedom which the experience of college sought to promote:

> Once chosen mother and daughter had to sit together at tea and supper for the whole year. If either of the two forgot and talked too much to her neighbour to the neglect of her duty, the other was jealous and meal times became periods of strife and boredom. The poor daughter often felt the position more acutely than her parent, for she was feeling horribly strange and had no one with whom she could discuss the position. (*Bishop Otter Magazine*, 1929)

The ritual of meal times, an essential component of domestic ideology, was translated into the college setting where, as in the home, it functioned not only to promote the socialization of individual students into college life but also as an important instrument of social control. Attendance at meals was compulsory and even as late as 1940 'If we students were late for a meal we had to ascend "On High" i.e. to the High Table where the staff sat, – and apologise' (HCA, Acc. no. 1404, 13, 1940). One student, twenty minutes late for supper, recorded:

> She told us that we had been 'reported out'. That means that the prefect at our table had reported us 'out' to the lecturer who was sitting on high. We were scared. We'd missed prayers and part of supper. We ran downstairs and the lecturer was just leaving. I hurriedly explained to her why we were late – we hadn't heard 'Big' – the bell – go and she said it was alright. (HCA, Acc. no. 888, 1928)

Cultural Horizons

A 'room of one's own' was one of the crucial experiences for the enhancement of feminine ideology which college could offer. Homerton students were fortunate in all having their own rooms, but at both Bishop Otter and Avery Hill dormitory accommodation divided into cubicles by flimsy wooden partitions or curtains remained the norm throughout the period. Both colleges accepted the desirability of giving students their own study bedrooms, but lack of funds held up the conversion programmes which were started in the 1920s; Bishop Otter did not vacate its last dormitory until 1957 (Shorney, 1989; Bishop Otter Annual Report 1927–28; *Bishop Otter Magazine*, 1958).

Homerton students were not only able to use their rooms for private study and to entertain fellow students, but they were also able to express their femininity by the individual arrangement of their possessions and treasures:

> My room is ever so sweet and I have just about the 'nicest' view in college. Our bureaus lock and we keep our valuables in them. . . . On my large window sill I have two small mats with a little photo on each . . . I have my clock on a larger mat on the mantelpiece and a blue box with snaps on one side and my scent spray on the other. (HCA, Acc. no. 887, 1928)

Students, however, were not allowed unfettered freedom to work or talk the night away. By 10 p.m. they had to call out to the corridor prefect 'In and Alone, Goodnight' (HCA, Acc. no. 1403, no. 4, 1935) and any student showing a light after 11 p.m. ran the risk of being noticed by the college porter on his nightly rounds and subsequently fined (*Homerton Roll News*, 1988–89).

The dormitory system at Bishop Otter offered little opportunity for either privacy or the expression of individual femininity. Cubicles had neither desk nor chairs – students were not allowed to study in them – and even dirty shoes had to be left downstairs (BOA, G78, 1920). The dormitory culture had, however, fervent supporters among students and even as late as 1958, when the system had been finally abolished, some 'bemoaned the free and friendly atmosphere of the "horse-box" in dormitories' (*Bishop Otter Magazine*, 1958). It could be that students who had never been away from home before found the camaraderie of the dormitory more comfortable than the lonely freedom of a room to themselves. Moreover dormitories were prominent in the culture of the upper-middle-class boarding school, and it is possible that lower-middle-class girls at training college, who had avidly read boarding school stories while at day school, saw the college dormitory system as an embodiment of their own fantasies (Cadogan and Craig, 1976).

The academic and cultural opportunities which the experience of college offered girls were widely appreciated – but there was an insularity about training college culture which, in spite of the possibilities for individual growth, militated against the development of a collective feminist consciousness. Significantly the two issues which did raise a collective response – peace and sexuality – are of continuing feminist concern today.

Many students testify to the permanent widening of their intellectual interests which was the result of the inspirational teaching which they received from college staff:

At Homerton I was lucky enough to be introduced to the eighteenth century mainly through the literature of that time, as directed by Miss Haynes. It was through her that I met James Woodforde (1740–1803) whose diary has been my bedside book for years. For that alone, as well as many happy memories, I have to thank Homerton. (*Homerton Roll News*, 1988–89)

Miss Allan had an enormous influence on me and I admired her tremendously. She was very keen on English and she used to say we must learn a lot of poetry and that has lasted me all my life. (*Homerton Roll News*, 1987–88)

Another student reflected the enduring values of her college education by continuing to record the books she had read, with comments, well into her old age (HCA, Acc. no. 1205).

Each college had a full selection of societies. At Avery Hill in the 1920s, 'there were art, debating, geographical, historical, literary, music and scientific societies' (Shorney, 1989), while at Homerton the college magazine in

1932 included reports from the Musical Society, the Student Christian Movement, Games Notes, Country Dancing Society, Guides, the League of Nations Union, and the Literary and Debating Society (*Homertonian*, 1932). Societies tended to wax and wane with individual enthusiasm. One wonders for instance at the commitment of the Homerton Debating Society in 1915 when students were urged 'Why not come to the meetings, if it is only to get your mending done' (*Homertonian*, 1915).

A feature of feminine culture peculiar to training colleges was the Saturday Evening Entertainment. Students were not normally allowed out of college in the evenings, except by special exeat, and on Saturday nights entertainment was provided in house. At Homerton 'concerts, one-act plays staged by students or "sedate all-female dances" ' (*Homerton Roll News*, 1987–88). No men of course were allowed into college on these occasions and 'the sedate all-female dances' evolved curious rituals which mimicked the courtship customs of the middle-class home. Students played the role of the young man while the staff played the girls whom they courted:

> Saturday evenings were best of all [when there was] dancing. The second year I had the privilege of taking a lecturer to the dance and called for her at her room. My favourite was Miss Glennie – she was very sweet. I always felt she had a soft spot for me and I certainly had for her. (HCA, Acc. no. 808, 1925)

Class hierarchies were also rigidly observed. The principal was sent a special printed invitation to the Saturday entertainment where she was escorted by the Senior Student (HCA, Acc. no. 1190, no. 44, 1927). 'She always wore evening dress' and received from her escort 'a knee rug, a cushion and sometimes a box of chocolates' (HCA, Acc. no. 1191, no. 6, 1932). At Bishop Otter a similar etiquette prevailed, and in the late 1920s it was considered an advance when at dances booking by programme was abolished and the 'grab' system installed in its place (*Bishop Otter Magazine*, 1926). Avery Hill on the other hand reflected its more progressive regime and access to metropolitan culture by allowing students as early as 1929 to attend a mixed NUT dance in the Woolwich Town Hall (Shorney, 1989).

By this time cultural opportunities outside the confines of college were becoming more readily available to students. At Homerton students had always been able to enjoy 'the beauty and delights of the Backs, the river and especially the unforgettable experience of Evensong at King's' (*Homerton Roll News*, 1987–88). But access to the academic and cultural facilities of the university itself was restricted. Only in the late 1920s were students, in the principal's words, allowed to 'receive a few crumbs which fall from the University table' (HCA, Acc. no. 1190, no. 3, 1927), and join the university branches of the League of Nations Union and the Student Christian Movement (HCA, Acc. no. 1190, no. 3, 1927–1929). By 1934 'fifteen students have had the privilege of singing in the chorus of CUMS [the University Musical Society]' (*Homerton Newsletter*, 1934). By the end of the 1930s when 'most of the lectures, both in the town and in the university are open to us' (*Homerton Newsletter*, 1938), a few bold students were active in the university branches of the Peace Pledge Union and the Labour Party (HCA, Acc. no.

1574). But Homerton students were still outsiders as far as the university was concerned. The 'titters' which students faced when they had to leave meetings early in order to get back to Homerton by 6 p.m. were an effective deterrent (HCA, Acc. no. 1031, 1932).

Cultural opportunities available at Bishop Otter were severely restricted by the college's provincial location, and it was not until the college was evacuated to Bromley during World War Two that students were able to enjoy the range of facilities available at Homerton or Avery Hill. The college had one memorable cultural highlight however in 1932, when John Gielgud, clad 'in a jaunty beret' and crying ' "a comb, a comb my Kingdom for a comb" ' arrived in his MG, and enthralled students with the 'beauty of his verse speaking' and his 'easy boyish manner' (*Bishop Otter Magazine*, 1932).

War or the threat of war formed the constant backdrop to everyone's experience at this time. Feminists of the period like Vera Brittain (Brittain, 1933) and Virginia Woolf (Woolf, 1938) forcibly articulated women's hatred of war, and their construction of military aggression as an essentially masculine discourse entirely inimicable to the feminine doctrine of peace. World War One, however, was a time of consolidation rather than catastrophe for the training colleges, who 'came through these four anxious years with wonderfully little inconvenience' (*Homerton Newsletter*, 1919). Only Avery Hill, with its vulnerability to air raids, the commandeering of one of its hostels for a military convalescent home, and the constant sight of the wounded in the London streets, really felt the impact of the war (Shorney, 1989). There were however isolated individuals in the colleges whose interest in the Women's Peace Movement prefigured the more general concern for peace which women students felt in the 1930s. Beatrice Collins, for instance, lecturer in art at Homerton, was one of the British women refused permission to attend the Women's Peace Congress at the Hague in 1915 (HCA, Acc. no. 1274), while a former Homerton student was given space in the college magazine in October 1915 for a long article on the International Peace Movement (*Homertonian*, 1915).

By the mid 1920s support for peace and disarmament was being increasingly articulated. In 1926 Avery Hill already had 100 members in its branch of the League of Nations Union, and in 1936 its always lively Debating Society carried a motion 'That under no circumstances is war justifiable' by a large majority (Shorney, 1989). At Homerton support for the peace movement was also strong in the 1930s. In 1932 students were addressed by Kathleen Courtney, Secretary of the Women's International League, on 'Disarmament and the World Crisis', and by the following year the college branch of the League of Nations Union had enrolled over sixty members (*Homertonian*, 1932; 1933). When the Peace Ballot was taken in college in the spring of 1935 'the result showed that Homerton is one hundred per cent pro-League'. In the same year an Anti-War Study Group was formed in college 'to bring an understanding of the war danger before students' (*Homertonian*, 1935). This desire for political activism was highly unusual for training college students and showed how strongly women felt about the issue of peace.

A great number of individuals are seeking out a means of producing peace merely by thinking about such a state. Active opposition against armaments, war propaganda and preparations is the only way. (*Homertonian*, 1935)

Four years later the peace movement had been overtaken by events and this time training colleges did not pass the war years unscathed. At Avery Hill no teaching at all was possible from the outbreak of war in September 1939 until the end of 1940, as fear of air raids and then the reality of the London Blitz forced the college to close its doors. Eventually Avery Hill was evacuated to Huddersfield but the lack of residential accommodation prevented a full college life (Shorney, 1989). Bishop Otter's buildings in Chichester were taken over by the Air Ministry at very short notice in September 1942 to provide accommodation for the planning of D-day. The college was obliged to move to Bromley, where it had to share premises with an Emergency Rest Centre (Bishop Otter Annual Report, 1941–46). These 'very difficult conditions' for the principal and her staff were however welcomed by students because it gave them access to the facilities of London and allowed them to escape from the provincial confines of Chichester (Bishop Otter Annual Report, 1941–46; *Bishop Otter Magazine*, 1944).

Homerton was more fortunate. Although it had to share its accommodation in the early part of the war with first Portsmouth and then Whitelands training colleges, by March 1942 the principal was able to announce that 'We now have Homerton to ourselves' (*Homerton Newsletter*, 1940–1942). Homerton's 'protected' wartime status seemed however only to encourage the insularity which had always been endemic among students, and to promote an emphasis on traditional feminine standards which had become impossible let alone obsolete in other more war-torn colleges. In May 1941, for instance, at a meeting of the newly formed Homerton Union of Students the 'interesting proposal' was put that 5s to 7s 6d per week for flowers 'is a scandalous extravagance when thousands are starving and homeless and that the money be diverted to the Red Cross'. The minutes recorded that 'College was in sympathy with the idea but regarded the expense as justifiable on the grounds that civilisation must be maintained amid the destruction of today' (HCA, Acc. no. 1241, 1941).

Similarly, proposals to allow students to wear trousers – slacks as they were then known – were twice defeated by the Homerton Union of Students, until finally, and after heated debate, a partial relaxation of this rule was passed in the autumn of 1943. Slacks could now be worn on cold winter evenings and on country expeditions, but they must be 'neat and well-fitting'. This restriction was all the more curious, and indicative of Homerton's insularity and old-fashioned standards, when, as had been pointed out during the debate, 'other women's colleges of both London and Cambridge universities' allowed the wearing of slacks without restriction (HCA, Acc. no. 1241, 1941 and 1943).

Sexual Issues

The issue of sexuality was bound to be crucial in institutions where marriageable young middle-class girls were under the care of a celibate female staff.

Moreover as sexual mores began to change, the needs and perceptions of college staff became increasingly opposed to those of the students. The principal and staff were concerned above all else to maintain the respectability of their institutions, and therefore to ensure that conditions were such that no breath of scandal could touch them; students, on the other hand, not only expected to have the same facilities for entertaining young men as they enjoyed in their own homes, but also looked forward to having the freedom at college to widen their male acquaintance.

Sensitivity to the possibility of scandal was particularly acute at Homerton. For Cambridge was a town dominated by the male values of its university which barely tolerated its own women students, let alone those in an obscure teacher training institution. Moreover at the beginning of our period it was less than twenty years since the university had lost its right to imprison women in the town – who could well have been Homerton students – whom it suspected of being prostitutes (Edwards, 1990). This overwhelming need to maintain the college's respectable image lay behind the principal's annual homily to students:

> Miss Allan gave us a lovely little lecture about the 5000 youths who are coming up in three weeks time. She says she's proud to say that Homerton girls have never behaved badly and she's sure that none of us will. She says if we've any friends [already] at the 'Varsity' we can go out to tea with them but we mustn't make friends. It's not done. (HCA, Acc. no. 888, 1928)

Allan's structuring of students' relationships with men in terms of respectability and social class rather than personal freedom or sexual choice was underlined by another student:

> My most vivid memory is of the lecture we were given on our first morning in college. We were to remember ... that we must never stand giggling with a man at the gate like a servant girl. (HCA, Acc. no. 1191, no. 17, 1931–1933)

Students' relationships with men friends were made as difficult as possible, although they were never actually forbidden. It was necessary to obtain 'written permission from parents to the Principal in order to meet an undergraduate in his rooms or his college. Afternoons only *and* accompanied by a college friend'. Men friends could be entertained in college – but only under very restricted conditions: 'to tea in a screened off corner of the drawing room' (HCA, Acc. no. 810, 1927). Even when a special room was eventually allocated for the purpose in the mid 1930s – when Alice Skillicorn, Mary Allan's successor as principal, was taking tentative steps to liberalize the Homerton regime – 'a lecturer popped in frequently to make sure you were behaving' (HCA, Acc. no. 1404, no. 13, 1940).

Interestingly, when a collective campaign to allow men to be entertained in students' rooms began in earnest in the 1940s, students used social reasons to argue for sexual freedom – the same tactics which the principal had earlier used to argue for the opposite case: 'It is difficult to entertain or talk to friends outside college ... there are not enough tea rooms' ... 'The point was made that no-one in College was able to return hospitality' (HCA, Acc. no. 1241, 1941, 1947).

The establishment of the Homerton Union of Students (HUS) in 1940 gave students the opportunity to use their collective strength to press for reform, and in October 1941 a proposition to allow outside 'friends' to be entertained in students' own rooms from 2 to 5 p.m. (HCA, Acc. no. 1241, 1941) was granted, but then, almost immediately the concession was withdrawn:

> During my two years, the Reps approached Miss Skillicorn on behalf of the students to ask for permission to entertain men friends in our rooms during certain times. The request was granted, but within a few weeks withdrawn at the request of students who found it disconcerting to find men upstairs when ... wearing dressing gowns. (*Homerton Roll News*, 1989–90)

For some students the presence of men in student's rooms was both an invasion of their own privacy, and an impediment to the corporate life of the women's community. This tension between individual desire for sexual expression and the corporate needs of a female community remained:

> Members of the University should be discouraged from treating Homerton as a place of entertainment or public house and should refrain from creating a disturbance outside after 'chucking out' at 10 p.m. (HCA, Acc. no. 1241, 1944)

Students tried to solve this dilemma by requesting better facilities for private entertaining in the college's public rooms, but this was refused. The principal avoided addressing the issue directly by couching her refusal in purely administrative terms:

> Miss Skillicorn saw the difficulties of entertaining in college and stated that it was not that she wished to prevent it but only to show that college as yet had not the facilities. (HCA, Acc. no. 1241, 1951)

Permission to entertain men friends in the privacy of students' rooms was not granted until the 1960s. Before then the only place where students could express their sexuality on college premises was in the semi-public conditions of what Homerton called 'the Back Drive'. This behaviour clearly offended against the middle-class standards of femininity which the whole college community upheld – but it was an illustration of the dilemmas which the whole issue of sexuality continued to pose.

> The Senior Student impressed upon the meeting that the question of discreet behaviour at the college gate had not been brought up by members of staff but by fellow students. It was felt very strongly that any behaviour at the gate could be noted by the residents in the district and it was up to students to see that their behaviour was not a bad reflexion on the College as a whole. She realised that Domestic

Staff were at times to blame but that did not mean that students need be indiscreet as well. (HCA, Acc. no. 1241, 1951)

Conclusion

In this chapter I have described and analyzed the culture of femininity which formed the enveloping context of the work of women's teacher training colleges. I have argued that, although many of the domestic and familial discourses of this culture were translated from the ideology of the middle-class home, nevertheless the ideology itself was crucially enhanced by the experience of college life. The cultural values of the liberal humanist tradition which were transmitted to students by college staff permanently enriched women's capacity for personal development and individual fulfilment; while the independence from family life widened women's horizons and gave them a freedom to experiment which would not have been possible in the middle-class home.

The training college culture with its combination of individual enrichment with collective stagnation is important to our understanding of the history of feminism in this period. Widening cultural and academic horizons permanently enhanced the lives of the individual women concerned, but the development of a feminist consciousness was by no means a necessary consequence of this enhancement. On the contrary, it can be argued that the very opposite was the case. Cultural enrichment not only upgraded women's skills as wives and mothers but it also supported them through the difficulties and frustrations of domestic life. Most crucially of all it distracted women from analyzing these frustrations into a collective feminist critique of the subordination of all women in a patriarchal culture.

Moreover the old-fashioned ethos and stagnant values of the training college world did not encourage students to seek collective solutions to individual grievances. In the early years of the twentieth century both staff and students had taken pride in the collective and pioneering values of their college communities. But by the 1930s the values of these feminine communities seemed increasingly outmoded to students, who sought instead for increased individual autonomy and sexual freedom. But it was not until the 1960s, when this autonomy and freedom had been largely achieved and the residential women's communities were no more, that paradoxically some women students began at last to seek a collective feminist response to their continuing subordination within a patriarchal culture.

The question remains: if the teaching profession was barred to married women, and if the surplus of women in the population[1] made marriage impossible for some, how far was the profession a real and positive choice for women? Evidence from Bishop Otter reveals that at the outbreak of World War Two about 50 per cent of former students were marrying within three to ten years of leaving college (Bishop Otter Register, 1939).

Summerfield has argued that the teaching profession was a positive choice for those girls, who had learned from the unmarried staff at their own schools that 'You didn't have to get married . . . you could make a good life without it' (Summerfield, 1987). Oram, on the other hand, has drawn attention to the hostility towards unmarried women teachers which was increasing from the 1930s onwards. This hostility was fuelled by economic factors and by changing attitudes to female sexuality. Spinster teachers and their relationships were becoming increasingly stigmatized and sexually marginalized in the wake of the *Well of Loneliness* trial in 1928. In the aftermath of this trial, which had given enormous publicity to the issue of lesbian relations, there was an increasing tendency to collapse the neutral label 'spinster' into the condemnatory one of 'lesbian'. Moreover the independence and social power of spinster teachers within their own communities also posed a real threat, not only to the economic position of their male colleagues, but also and more generally to the prevailing discourse of women's subordination to men within heterosexual marriage (Oram, 1989).

I would argue however, drawing on both Summerfield and Oram's perspectives, that the training college experience was so popular just because it did allow girls to 'hedge their bets'; some women would teach until their retirement on marriage, while those who, either from choice or necessity, did not marry, were well equipped, in spite of society's negative attitude to unmarried women, to maintain their independence by pursuing a satisfying career.

Note

1 In the early 1920s most local education authorities introduced regulations requiring women teachers to resign on marriage and this marriage bar remained in force until World War Two (Oram, 1989). Moreover, as a result of the loss of men during World War One, 'the number of "excess" women to men in the population increased from 664,000 in 1911 to 1,174,000 in 1921 and still remained at 842,000 in 1931' (Beddoe, 1989).

References

BEDDOE, D. (1989) *Back to Home and Duty: Women between the Wars, 1918–1939*, London, Pandora.

BISHOP OTTER ANNUAL REPORTS 1925–26; 1927–28; 1941–46.

BISHOP OTTER ARCHIVE (BOA), G78: Reminiscences of a student 1920–1922.

BISHOP OTTER MAGAZINE 1922; 1926; 1929; 1932; 1944; 1958; 1971.

BISHOP OTTER REGISTER 1939.

BRITTAIN, V. (1933) *Testament of Youth*, London, Gollancz.

CADOGAN, M. and CRAIG, P. (1976) *You're a Brick, Angela!*, London, Gollancz.

EDWARDS, E. (1990) 'Educational Institutions or Extended Families? The Reconstruction of Gender in Women's Colleges in the Late 19th and Early 20th Centuries', *Gender and Education*, **2**, 1, pp. 17–35.

EDWARDS, E. (1992) 'Alice Havergal Skillicorn, Principal of Homerton College, Cambridge 1935–1960: A Study of Gender and Power', *Women's History Review*, **1**, 1, pp. 109–29.

HOMERTON COLLEGE ARCHIVE (HCA) Acc. no. 807: Reminiscences of a student 1922–1925.

HOMERTON COLLEGE ARCHIVE (HCA) Acc. no. 808: Reminiscences of a student 1925–1927.

HOMERTON COLLEGE ARCHIVE (HCA) Acc. no. 810: Reminiscences of a student 1927–1929.

HOMERTON COLLEGE ARCHIVE (HCA) Acc. nos. 887, 888, 889, and 895: Letters home from a student 1928–1930.

HOMERTON COLLEGE ARCHIVE (HCA) Acc. no. 1031: Interview with two students 1932–1934.

HOMERTON COLLEGE ARCHIVE (HCA) Acc. no. 1190, nos. 3 and 44: Replies to questionnaire on Miss Allan from two students 1927–1929.

HOMERTON COLLEGE ARCHIVE (HCA) Acc. no. 1191, nos. 6 and 17: Replies to questionnaire on Miss Allan from two students 1932–1934 and student 1931–1933.

HOMERTON COLLEGE ARCHIVE (HCA) Acc. no. 1205, Extracts from commonplace book, 1971, belonging to a student 1913–1915.

HOMERTON COLLEGE ARCHIVE (HCA) Acc. no. 1274, 1915.

HOMERTON COLLEGE ARCHIVE (HCA) Acc. no. 1241: Minute and record book of Homerton Union of Students, 1940–1954.

HOMERTON COLLEGE ARCHIVE (HCA) Acc. no. 1403, nos. 3 and 4: Replies to questionnaire on Miss Skillicorn from students 1935–1937.

HOMERTON COLLEGE ARCHIVE (HCA) Acc. no. 1404, no. 13: Reply to questionnaire on Miss Skillicorn from student 1940–1942.

HOMERTON COLLEGE ARCHIVE (HCA) Acc. no. 1412: Interview with a member of staff 1942–1946.

HOMERTON COLLEGE ARCHIVE (HCA) Acc. no. 1574: Interview with two students 1938–1940.

HOMERTON COLLEGE REGISTERS 1914; 1924; 1929; 1939.

HOMERTON NEWSLETTER, 1919; 1934; 1938 and 1940; 1941 and 1942.

HOMERTON ROLL NEWS, 1987–88; 1988–89; 1989–90.

HOMERTONIAN, 1915; 1932; 1933; 1935.

ORAM, A. (1989) ' "Embittered, Sexless or Homosexual": Attacks on Spinster Teachers 1918–39', in LESBIAN HISTORY GROUP, *Not a Passing Phase: Reclaiming Lesbians in History 1840–1985*, London, The Women's Press, pp. 99–118.

SHORNEY, D. (1989) *Teachers in Training 190–1985: A History of Avery Hill College*, London, Thames Polytechnic.

SUMMERFIELD, P. (1987) 'Cultural Reproduction in the Education of Girls: A Study of Girls' Secondary Schooling in Two Lancashire Towns 1900–1950', in HUNT, F. (Ed.) *Lessons for Life: The Schooling of Girls and Women 1850–1950*, Oxford, Blackwell, pp. 149–70.

WIDDOWSON, F. (1983) *Going Up into the Next Class: Women and Elementary Teacher Training*, London, Hutchinson.

WOOLF, V. (1938) *Three Guineas*, London, Hogarth.

Chapter 6

The Diary of Doreen Bates:
Single Parenthood and the Civil Service

Elizabeth McClair

In the summer of 1990 I began looking at the diaries held in the Mass Obser-vation Archive, University of Sussex Library. My interest focused mainly on those written by women and I became immersed in a spectrum of extracts that were as diverse and unpredictable as they could be. I was fascinated. Here, I felt, was a history written by women in a way that gave the reader an insight and a sense of authenticity that was quite unique. Women were defining their lives. Few saw themselves as writers, that was not their aim. They had been asked to relate the details of everyday life, and so, perhaps for the first time, their daily routines and chores took on, publicly, the impor-tance they held privately.

As I read more and more, the voices of these women from the past began to take on a significance that I believed was important. Some wrote extensively, others just a few lines jotted down hurriedly at the end of each day. Some wrote about their personal lives whilst others kept strictly to factual information relating to food shortages, evacuees, or other matters pertaining to wartime conditions. Among the voices was a medley of opinions, ideas, and beliefs across a broad band of class and politics. Each diary, in its own individual approach, reflected the way in which women's lives functioned within the society of that time.

Soon I became familiar with some of the more prolific diarists. I began to be caught up in the stories they revealed about their lives. Underlying these stories lay the conditions that shaped them. Often I came across a deep-seated anger or dissatisfaction, sometimes voiced, sometimes merely intimated, with the unjustness and inequality of these conditions. For some the war had brought an opportunity to break out of the confines of the dom-estic sphere into a more public arena, whilst others found that the material circumstances of shortages made their lives considerably more difficult as they struggled to provide their families with adequate food and clothing.

As a picture of the wartime years built up, as it were from the inside, from the lives of these women as they related it daily, I began to see that it was from the way women went about their day to day lives that any kind of resistance to inequality could effectively be instigated.

I began to search for one diary that would convey this in the way that the diaries collectively conveyed it. My undertaking seemed an impossible one among the myriad of extracts, and with only a vague and instinctive idea of what I was looking for I continued to read. When eventually I chanced upon an extract from the diary of Doreen Bates I felt that my search might have ended.

As I pieced together her diary over its three-and-a-half-year period I realized that this was an important documentation both in terms of the diarist's own life and in its relevance to the lives of other women. Doreen was middle-class and educated but what she attempted to achieve in the period the diary covers was to have repercussions certainly within her own class and to a lesser extent in the lives of working-class women too.

The Diary

Doreen Bates' diary opens in September 1940 when she was a single woman aged 32 and a Tax Inspector in the Civil Service. She stated that the purpose of the diary was to record, for Mass Observation, the details of everyday life. The first entries did this meticulously, describing air raids and the damage caused by them, the difficulties of travelling to and from work and the endless delays of trains and buses. Personal details are scant, but gradually a picture emerges and the reader learns that the diarist lived with her widowed mother and sister, had a love of music and the theatre, and was involved with a man referred to as 'E'.

By January 1941 more facts about the diarist's life creep in and the reader begins to suspect from the unusual times and places of their meetings – a walk on Wimbledon Common early in the morning, fleeting lunches, and hurried assignations at Victoria Station before work – that 'E' was married. From the entry on 1 February 1941 it becomes clear that her relationship with him had become an intimate one: 'Had the morning off with 'E' in Elsie's room'. On the same day she learned that she was to be transferred to Belfast. The news was distressing to her and she attempted, unsuccesfully, to oppose it.

In March 1941 she took up residence in Ireland. It is at this point that it becomes apparent to the reader that she was pregnant, a fact that she regarded with quiet delight and a certain humour: 'I begin to bulge noticeably'.

On 29 May 1941 she returned to London for 'an emotional weekend' in which she met up with 'E', the child's father, broke the news to her devastated mother, and attempted to negotiate 'sick leave' for her confinement.

On her return to Belfast, and with the facts brought out into the open, as far as her direct superiors within the Civil Service were concerned, she was relegated to working at home for the duration of her pregnancy. Whilst there was some sympathy for her among her colleagues it was not thought suitable for her, in the light of her rank, to appear in the public workplace. At the same time her right to paid leave during her confinement became an issue:

August 1941
He heard from B (Director of Establishments) on Friday that the Treasury have decided on sick leave without pay. I heard the story.
A woman clerk is having a baby (unwillingly) and went to her

association for advice. The Secretary sent her to Miss H, the woman establishment officer (appointed solely to protect and watch the interests of the women in the department!). She said sick without pay. Secretary wasn't satisfied and went to B who said nonsense, sick leave with pay – we have just decided on the 'hypothetical case' (which is me!). Miss H got on her high horses, went to the Treasury which after long consideration supported her. B is cross, partly on general grounds (he is sympathetic) and partly because he has lost face! It is clear that no more can be done inside the department. McC was nice and said he could probably arrange for a loan from the Benevolent Fund if necessary!

Later that month she returned to London and encountered more difficulties, both emotional and practical, which she approached with a dry humour: 'I rang up home today and R [her mother] invited me for next weekend – from after dark on Saturday to after dark on Sunday!' This sense of humour carried her through the more exasperating difficulties she encountered in attempting to book herself into a hospital for the birth of her child:

Queen Charlotte's was more communicative after Dr M's note. I liked the doctor who examined me very thoroughly and said all was well. Later I saw him again and he said Dr M was an old friend of his and he would admit me if he could but the Government strictly limited their numbers and he couldn't do anything now. He asked me what my job was and whether I was having the baby adopted! When I said no, I wanted it, he said 'Oh I see, planned for was it?' The almoner phoned St Mary's Paddington and made me an appointment. I saw their almoner who was younger and more business like but I didn't like the place. It was the old part (1897) and semi-basement. They could take me till in filling up the forms we came to husband's name. There I learnt that they weren't admitting unmarried women apparently because they were limited in numbers and so discriminated. She would put me in touch with someone who might help. I had a letter from the Society of Moral Welfare! I was inclined to see what it would be like but M [her sister] said 'Don't waste their time – it is a difficult job.' Meantime I have got into touch with the Stanmore Nursing Home and saw the Matron yesterday. I liked the place and it will be better if the blitz begins.

In October 1941 to her great delight Doreen Bates gave birth to not one child but two. By this time the diary had entirely broken from its original concept of factual documentation and become a highly personal recording of events:

I didn't want a hot bath or an enema as the pains were very frequent. I watched the clock above the bed and tried to relax between the pains and 'push down' when they began. Sister said later from my size they expected a 5 or 6 lb baby and no trouble at all. She told me the baby ought to be born by 10.00 but at 9.30 the pains decreased in frequency and intensity. She phoned for Dr B who appeared in a large orange rubber apron. He gave me two tablets (I think bromide). My back was aching all the time so that I could barely distinguish the pains and I seemed to have very little energy to push. Doctor said the baby's head was only $1\frac{1}{2}$ inches in and another one or two 'good strong pains' would bring it far enough for me to have chloroform for the actual delivery. But the pains slowed up at about 10.45 and he put a dark cone over my face. I smelled the sickly sweet chloroform and tho I had been longing for it I pushed it away with my hand. Dark clouds like smoke seemed to roll up and extinguish everything. At 1.00 am Sister's voice urgent and summoning – seemed to come from an immense distance waking me up. I said 'Is the baby alright?' – She said 'Yes – you've got twins! A boy and a girl!'

After the excitement of the birth the diarist settled down to the practicalities of bringing up her two children alone. Her job was of the utmost importance to her and her next step was to look for childcare for the twins. She rented a house, employed a nanny, and, six weeks after the birth, she returned to work.

Saw McC to arrange to return to work next Monday. He gave me a message from B that if I repeated my performance I should be sacked. McC said he told B that he would deliver the message but the Association would protect against 'such unduly harsh treatment' and he felt B himself would not go so far if the occasion arose. McC seemed impressed to see me looking so fit and remarked that maternity had not changed me at all. I left him to lunch with E having arranged to call at Kensington to collect from M my cheque and the voucher for my watch which has been repaired. I collected it at Walkers, Victoria. I then went to Fulham to look at a second hand twin pram (12–14 guineas) in a depository. I wasn't sure about its condition – especially the tyres, so I thought it over.

Under Civil Service ruling at that time a woman, on marriage, was obliged to resign her job and the question arose as to whether the diarist was to be regarded as a married woman and treated accordingly. She maintained that as a sole earner with two children to keep (also a mother whose household she appeared to be contributing heavily to) she should be regarded as any other 'breadwinner', that is, as a man, with all the rights and privileges that go with that position.

Doreen Bates' fight continued over a period of two and a half years and whilst she maintained her position in the Civil Service the argument she engendered was still inconclusive when the diary was discontinued in 1944. During these years the diary is taken up primarily with domestic details concerning her two growing children and her relationship with their father whose love for his new 'family' did not induce him to leave his wife although she was informed of the situation and did meet the diarist and the twins. The diarist reveals herself to be content, happy, and fulfilled and the reader is left with the deep impression of an independent and strong woman; one who quietly refuses to be relegated to the marginal position in which her desire for motherhood whilst unmarried would normally place her. It becomes clear from a note inserted in an earlier section of the diary that she was not un-aware of the importance, historically and politically, of her position. She wrote:

> The diary strikes me as having very little war diary. But I left in a certain amount of my personal affairs – mainly because they are probably of general interest at the moment. I believe I am the first woman civil servant of my rank to challenge the marriage law by pushing it to its logical conclusion – dispensing with the ceremony and starting a baby.

Doreen Bates' diary is, I believe, an important documentation of the way in which political struggle is fought out in everyday life. Women such as Doreen have contributed in a significant way to the fight against oppression by challenging the institutions and policies that inhibit change. Her diary is an impressive record of that fight. Doreen Bates' contemporary counterpart, Janice Poole, an RAF sergeant and unmarried mother of twins, is currently running a campaign to win the right to married quarters. The official ruling allowing women to stay on in the Services after having children was only changed in 1990. History moves slowly.

Section II

Political History

Chapter 7

Gendering Patriotism: Emmeline and Christabel Pankhurst and World War One

Jacqueline de Vries

When war broke out in August 1914, Emmeline and Christabel Pankhurst, the leaders of the militant Women's Social and Political Union (WSPU), hailed it as the great apocalypse. The war was the logical fulfilment, they reasoned, of suffragette prophecies of imminent world disaster consequent on male domination and vice. European governments had rejected the aid of women and had thereby brought the war upon themselves. Mrs Pankhurst and her daughter were in St Malo, France, when Great Britain announced its entrance into the conflict, and Christabel immediately drafted her ideas into an article to send to London for publication in the 7 August edition of the *Suffragette*:

> A dreadful war-cloud seems about to burst and deluge the peoples of Europe with fire, slaughter, ruin – this then is the World as men have made it, life as men have ordered it. A man-made civilization, hideous and cruel enough in time of peace, is to be destroyed.... This great war ... is Nature's vengeance – is God's vengeance upon the people who held women in subjection, and by doing that have destroyed the perfect human balance.... Only by the help of women as citizens can the World be saved.... Women of the WSPU, we must protect our Union through everything. It has great tasks to perform, it has much to do for the saving of humanity.[1]

This great war, Christabel argued, signalled the demise of an era of male domination. It was her belief that a 'woman's era' would rise in its ashes.[2]

Despite Christabel's firm and determined tone in this editorial, confusion reigned among the leaders of the WSPU, who, like the leaders of other suffrage organizations, were struggling to find an appropriate course of action in the face of this Armageddon.[3] Their decision-making during those early days of war was hastened by the chaos of external events. On 10 August, Home Secretary Reginald McKenna cancelled all prison sentences for suffragettes, most of whom were languishing in Holloway under brutal, self-imposed hunger strikes. During the preceding months most of the WSPU's energies had been focused on promoting the image of a male government cruelly torturing its women, but McKenna's announcement ended the most visible persecution of women and eliminated one of the WSPU's most effective rallying points. Three days later, Emmeline Pankhurst issued a statement to WSPU members temporarily suspending militancy. The war, Mrs Pankhurst asserted, was itself proof enough that the present system of governing was in dire need of revision:

> Even the outbreak of war could not affect the action of the WSPU as long as our comrades were in prison and under torture.... It is obvious that even the most vigorous militancy of the WSPU is for the time being rendered less effective by contrast with the infinitely greater violence done in the present war not to mere property and economic prosperity alone, but to human life ... and with that *patriotism* which has nerved women to endure endless torture in prison cells for the national good, we ardently desire that our country shall be victorious.[4]

After months of guerilla warfare against the British government, the suffragettes were viewed by many observers as a sort of fifth column, hardly capable of love and support for their country. But Mrs Pankhurst's depiction here of suffragettes as 'patriotic' is more than a shallow manipulation of wartime discourses. Rather, she is emphasizing a longtime theme of WSPU agitation: that the suffragettes' arduous battle for the vote was part of a wider struggle toward a national regeneration based on principles of self-sacrifice.[5] The outbreak of hostilities underscored suffragette arguments that nations would continue to degenerate until women were granted power, in the form of the vote, to influence their nation's policies.[6]

Whether militant or not, suffrage workers recognized that their wartime behaviour would help determine their political future. The war, they reasoned, offered an unprecedented occasion to demonstrate the women's vital role in safeguarding the future of the British nation. But, as was the case during the pre-war campaign, suffrage workers continued to disagree about the means to this end. Many feminists believed – or came to accept the idea – that their primary purpose was to work for peace.[7] In contrast, the WSPU, under Christabel and Emmeline's direction, cast its enormous energies on the side of the war effort. The vastly different approaches to the war among Britain's various feminist organizations not only fragmented the movement so severely that it would never regain its pre-war coherence, but also signalled a gradual shift in orientation: from a feminism premised upon late-nineteenth-century national ideologies which focused on themes of regeneration to a feminism which sought international alliances and solutions and emphasized women's individual needs.

Scholars have written little about patriotic women during World War One, and the accounts that do exist are often cursory and give far more attention to the minority of women who became pacifists during the war.[8] Some historians have portrayed the Pankhursts' jingoism as an about-face and an abrupt rejection of the WSPU's pre-war policies,[9] but this assessment fails to delve beneath surface events or consider the crucial relationship between feminism and national ideology. Other historians have probed more deeply and argued that the Pankhursts' chauvinism logically followed from their pre-war militancy,[10] but then go on to suggest that their vehement patriotism signified an acquiescence to male cultural ideology.[11] While provocative, this assessment discounts the many gendered aspects of the Pankhursts' war-time discourse and overlooks important ideological continuities from before the war. This approach also highlights what may be a subtle unwillingness among some feminist historians to acknowledge feminism's complicity with the destructive, 'masculine' act of war.[12]

The Pankhursts' support for the British war effort stemmed from a complex mix of patriotic discourses, feminist ideas, and personal styles. Close examination of the Pankhursts' jingoism reveals an ongoing dedication to the ideals of purity, sacrifice, and national moral regeneration, and to a belief in women's unique capability to achieve these goals. The Pankhursts couched their decision to take up anti-German banners within a *gendered* interpretation of the war, in which they defined Germany as a masculine nation and Britain and her allies as feminine. Such a construction, while seemingly at odds with the rhetoric of their earlier campaigns, allowed them to support Britain's involvement in the war while maintaining the basic tenets of their pre-war feminist perspective.

In re-examining the Pankhursts' patriotic discourses during the war, this essay highlights a theme which has been developed by the work of many others: that the ideology we call 'feminism' is a product of, and must be understood within, the specificities of historical time and place. As Antoinette Burton has recently argued, feminists are thoroughly embedded 'in their own cultures, national assumptions, and geopolitical formations.'[13] It may be tempting to conclude that the Pankhursts, and other feminists who failed to take a clear stand against the war, were in this case 'British first, women second'. But the rather complicated reality was that British feminists 'occupied a place at the crossroads of several interlocking identities'.[14] The Pankhursts were simultaneously feminists *and* British, with one identity infusing the other. Although their patriotism may often have been expressed in a manner similar to that of, for example, the press mogul Lord Northcliffe, we shall see that it was informed by a different, arguably feminist, set of ideals. It is to these issues – the Pankhursts' discourse on the nation, womanhood, and patriotism – that I now turn.

Patriotic Suffragettes

After Mrs Pankhurst's demobilization declaration, while other WSPU members recuperated from the effects of hunger strikes, police brutality, and months of endless meetings, Christabel Pankhurst embarked in the autumn of 1914 on a speaking tour through the United States and Canada to raise support for Britain's war effort. Significantly, these public forums gave Christabel the opportunity to represent her country as morally just and to introduce herself as both a feminist and an envoy of the British nation. In a typical presentation on 24 October, Christabel skillfully interwove themes of patriotism, religious duty, and gender equality, which were to remain tenets of WSPU propaganda for the next four years.[15]

The war, Christabel argued in front of a large crowd in New York City's Carnegie Hall, was a manifestation of the ongoing struggle between moral force (as advanced by the suffragettes) and physical force. In her opinion, Germany held as its 'national religion the theory that might is right', and aimed to dominate all of Europe by means of physical force rather than the moral force of justice. All this, she reasoned, resulted from Germany's glorification of masculinity:

> Bismarck boasted that Germany is a male nation. We do not want male nations. 'Male and female He created them!' A nation exclusively dominated by the male and by the ideas of the male is a nation governed unnaturally; is a nation which is bound to go wrong. The more you have the man's and the woman's point of view balanced, the more sane will be the nation, the more just and wise will be the nation.[16]

She conceded that her own country was not the perfect example of democracy and gender equality, which of course was why suffragettes had been campaigning so vehemently before the war. Reaffirming her commitment to suffrage militancy, she explained, '[this] is not to say that I prefer war to peace; but it is to say that when people want to govern me by physical force and not by the moral force of justice then I am prepared to defy their physical force to the very death'.[17] Germany's invasion of Belgium was clearly an exercise of pure physical force, motivated by forces more sinister than those within her own country. At least Britain, she suggested, currently rejected the doctrine of 'might makes right' and had been pulled into the war out of principles of duty and honour rather than impulses of naked aggression.[18]

What was more, Christabel argued, Britain was a liberal, Christian democracy which was acting out the national ideals set forth by Giuseppe Mazzini, the great Italian patriot and leader of Italy's *risorgimento*, who had long served as an inspiration to English feminists.[19] Quoting from Mazzini's 'The Duties of Man', Christabel charged: ' "What can each of you do with your isolated powers for the common improvement, for the progress of humanity? . . . God gave you the means you want when he gave you a country. . . . The love of country is the right preparation for love of humanity as a whole" '.[20] She continued to stress throughout her speech Mazzini's ideal of patriotism as the means for completing God's mission to humanity:

> We cannot love other countries, we cannot understand them unless we first understand and love our own country. Some people seem to think that patriotism is out of date. I do not believe that. It is true we do not want a selfish patriotism. We do not want a greedy patriotism. We want a wise and great hearted patriotism. We must agree with Mazzini when he says: 'Your country is the token of the mission which God has given you to fulfill to humanity.'[21]

Within this romantic patriotic ideal, women had a special role – a role which feminists had claimed long before the outbreak of war. This vision was best expressed in one of Mazzini's most famous quotes, which appeared on suffrage stationery and publications: 'Seek in women not merely a comfort, but a force, an inspiration, the redoubling of your intellectual and moral faculties. . . . Consider her your equal in civil and political life'. Thus, Christabel believed that suffragettes should view the war as an invitation to show themselves as the very life force of the British nation – a nation which was a living organism, to be strengthened and protected, so that it would promote the good of all humanity.

As Christabel concluded her speech, she warded off potential accusations that the suffragettes had been unpatriotic in their campaign against the British government for the right to vote. Such an assessment stemmed from a superficial understanding of their movement: 'Good heavens! Why should we fight for British citizenship if we do not most highly prize it?' The suffragettes' battle had never been against the nation, but rather against the infectious forces corrupting the nation. She ended by saying that it was women to whom nations must look for a peaceful future.

Back in England, WSPU members were also struggling to understand the relationship between feminism and patriotism. The WSPU's main leaders before the war – Mrs Flora Drummond, Annie Kenney, Grace Roe, and Mrs Dacre Fox – apparently accepted Christabel's ideas and all continued to hold prominent positions in the organization throughout the war. Approval of the leaders' pro-war stance among the rank and file WSPU membership is difficult to chart, as there are no available membership records and few financial reports for this period. The organization retained substantial funds collected prior to 1 August 1914, and continued to maintain its palatial central offices in Lincoln's Inn House, Kingsway, and a sizeable staff until late 1915. At that time the WSPU moved to more modest offices on Great Portland Street, a transition common to nearly every other suffrage organization facing a decline in membership support due to other pressing wartime financial concerns. Donations to the WSPU appear to have dropped, although it is unclear whether the cause was the WSPU's support for the war or the economic hardship brought on by the war economy.[22] What is clear, though, is that at least two factions split from the main group to pursue alternative strategies.

The Suffragettes of the WSPU (SWSPU) formed in October 1915 after some WSPU and ex-WSPU members met to protest the Pankhursts' use of the WSPU's funds and name in pursuit of issues not directly related to women's enfranchisement. Members of this new organization seemed to agree with Christabel that European governments' neglect of 'women's point of view' was a major cause of the war, but they believed that campaigning for the vote, and not war recruitment, was the most effective way to demonstrate their patriotism, strengthen their country, and prevent future wars. Although their monthly publication, the *Suffragette News Sheet*[23] occasionally printed direct criticism of the WSPU, some of the SWSPU renegades maintained ties of friendship with Christabel and the other leaders and gave tacit support to their activities. In March 1916, another group of WSPU members formed a second offshoot organization under the name of Independent WSPU (IWSPU).[24] Its journal, *The Independent Suffragette*, stated that the group intended to 'work in the spirit of the old WSPU . . . to unite members for suffrage work', but the journal indicates that this group apparently was more hostile to the Pankhursts. Neither organization, however, appeared to be openly pacifist, but rather followed a policy of avoiding an official stand on the war issue which might alienate segments of their membership.

Representations of Patriotic Womanhood

After an eight-month hiatus, publication of *The Suffragette* resumed in April 1915. Although its form and layout was nearly identical to the former version, the headlines and articles seemed to have little to do with women and feminism, focusing far more on mobilization efforts. Yet an editorial in this first issue revealed a distinct ideological continuity from before the war:

> What self-respect and dignity are to the individual, the patriotism of its members is to the Nation. This paper, *The Suffragette*, has always sought to rouse women to a sense of their personal dignity and importance, and of their rights as individuals, and so quite naturally and logically, in the present national crisis, our appeal is to the patriotism of women. In militant women, the love of country is necessarily strong. The supreme reason why we have fought for the vote is that we might obtain the power to help in making British civilization an even finer contribution to the civilization of the world than it has been in the past.[25]

The editorialist, probably Christabel, clearly has not abandoned her desire for the vote. Rather, as was the case before the war, she viewed the vote not as an end in itself but as a means to elevate humanity. In this passage Christabel draws a parallel between suffragism and patriotism: both aimed to heighten women's dignity and self-respect and thereby uplift the whole nation. Her comparison is premised upon the belief that women represent the nation; indeed, that they embody the true meaning of the nation.

Another article in the 16 April *Suffragette*, entitled 'Our Country is Our Temple', attributes a spiritual element to women's patriotic activities. Again, Mazzini provides the inspiration:

> He who seeks by force to destroy [the nation], is our enemy, and we must resist him, having as our warrant and justification for such resistance the example of Christ, who drove out the money-changers from the temple, with a whip of small cords.[26]

Just as suffragettes had withstood attacks on their bodies of forcible feeding and police violence, so now they must help the nation withstand the encroaching evils of German militarism.

Nationalist patriotism, like militant suffragism, became a kind of religion for Christabel and her followers. They believed that individual desires and convictions must be sacrificed to the work of securing and purifying the British nation. As Mrs Drummond wrote in 1916, 'Nationalism [is] one of the noblest principles of human life – the true corrective of individual selfishness, the basis of *real* internationalism'.[27] Likewise Christabel declared that the first task in the struggle to win the war was to 'subordinate the personal interests and the *amour propre* of the individual for the good of the whole nation'.[28] But through such selfless faith, the nation would be preserved and the individual ultimately exalted.

During the war, then, the WSPU's leaders believed that their role was to inspire women to protect the British nation from any force, both within and without, that threatened its existence. Their most immediate fear was of German infiltration of the home front and their objective became the purification of England of any naturalized Germans, English of German descent, and even 'any British blood who may be pro-German or half-hearted in the prosecution of the war'.[29] While such chauvinism resembled the popular anti-German discourses circulating through newspapers and recruiting meetings, it arose from a distinctly gendered analysis. Whereas national public opinion tended to blame the 'German peril' on the Prussian military system or Germany's lack of democracy, the Pankhursts saw these as only symptoms of the real problem, which they located in Germany's unbalanced gender system. At one of the first of the WSPU's weekly war service meetings, held on Thursday afternoons in Piccadilly Circus, Mrs Pankhurst castigated Germany's gender inequities:

> Because Prussianisation is masculinity carried to a point of enormity and obscenity even, that is what we women are fighting against in every land and in every race, we are fighting against that over sexuality [*sic*] that we women have always been trying to break down.[30]

The reputation of German males for 'over sexuality' was linked to reports of their allegedly perverse 'acts of mutilation' on the battlefield.[31] Editorials and articles in *Britannia* (significantly, in October 1915 the journal was renamed *Britannia*, a traditional symbol of the triumph of English virtue over foreign vice)[32] voiced the fear that such deviant behaviour was insidious and might infect English society and weaken its strong 'moral fibre'. A triumphant Germany, Christabel argued, would set the women's movement back a hundred years: 'If a German victory would be an appalling calamity for men, it would for women be infinitely worse. To defeat the Germans is the Woman question of the present time'.[33]

The Pankhursts viewed the war with Germany as evidence of the ongoing 'sex war,' which would end only when femininity had overcome masculinity's unjust dictatorship. Each country involved in the conflict, they believed, embodied either masculine or feminine values; of course their sympathy lay with the 'feminine' nations of France, Belgium, and Serbia, personified in *Britannia*'s illustrations as humble, yet determined peasant women. Belgium in particular was regarded as the 'suffragette country', brutalized and violated by German masculinism and militarism.[34] Just as the men in a 'feminine' nation were good, the women of a 'masculine' nation were the reverse.[35]

The 'feminine' representations of certain nations and their peoples in *Britannia* reveal the Pankhursts' own radical vision of a thoroughly gendered political sphere, in which women's interests were implicitly tied to national politics and international diplomacy. One poignant example can be found in *Britannia* during the autumn of 1915, when Austro-Hungarian troops were threatening to overrun Serbia and Allied military strategy seemed ineffective. The journal took up the Serbian cause with a vengeance, paralleling Serbia's efforts with the suffragettes' own struggle to achieve political power and improve the human condition. A deluge of articles explored their national history, character, and lifestyle. Cast as moral and courageous freedom-fighters, the Serbs were described as being engaged in a long battle 'for Liberty and [to] defend Christendom', first against the 'Ottoman yoke' and now against German domination. Animated by a righteous spirit, Serbians had not passively allowed themselves to be conquered, but valiantly revolted, fighting 'so bravely not for possessions, not for material gain, but in order to preserve liberty and independence'.[36] Serbia's femininity, then, was recognizable by its courage and spiritual vision to fight for liberty and freedom in the face of nearly insurmountable opposition.

This particular conception of patriotic femininity was also apparent in the WSPU's use of the images of Joan of Arc, a popular suffragette heroine before the war, along with a new role model – the Russian Madame Botchkareva.[37] The Pankhursts favoured these women over such figures as Florence Nightingale, who represented the more traditional female virtues of a nurturing maternalism, because they embodied physical courage, spiritual tenacity, and devotion to the nation. According to legend, both Joan of Arc and Madame Botchkareva had been warriors for their country and had bravely persisted after even the strongest men had failed. When France faced imminent defeat at the battle of Orleans in 1429, Joan of Arc dutifully obeyed the voice of God, who told her to assume the head of France's forces and lead them to a glorious victory. Similarly, in early 1917, after only five weeks of training, Madame Botchkareva had led her women's battalion – the Regiment of Women to the Death – to the war front to replace retreating (male) Russian troops. Almost immediately they captured a hundred prisoners of war, including two German officers.

Britannia featured Botchkareva and her tough women soldiers as modern-day versions of Joan of Arc. Labelled 'The Women's Example', her picture confronted readers of the 3 August 1917 issue of *Britannia*. Her arms folded in determination, Botchkareva stared emotionlessly, challenging the camera's eye. With strong, sexless facial features, and dressed in a man's army uniform decorated with medals of honor, her image seemed to subvert every popular conception of beauty and femininity. But like the young Joan, Botchkareva represented the newly defined feminine ideals of selflessness, courage, and devotion to national freedom. In positing Botchkareva as British women's role model, Christabel deliberately challenged the notion that femininity could be found only in physical attractivenes, emotional weakness, and the biological function of motherhood. Her understanding of femininity transcended a biological essentialism and centred on what one might call a 'spiritual essentialism' – a belief in women's unique, spiritual relationship to the nation. It was not just a woman's maternal qualities, but also her determined, courageous, and patriotic stand for what was 'right' – for freedom, democracy, and morality – that necessitated her immediate and full participation in the nation's political and diplomatic functions.

Visions of the Future

During the summer of 1917, while Madame Botchkareva prevailed over Russia's enemies, the British House of Commons passed a Bill to enfranchise soldiers, sailors, and women over the age of 30.[38] In November 1917, when passage of the Bill seemed assured, the remaining WSPU leadership, under pressure from SWSPU and IWSPU, renamed their organization 'The Women's Party'.[39] Rather than integrating into one of the established political parties – Conservatives, Liberals, or Labour – they were determined to retain their independent voice and attract a constituency of women only. In an inaugural speech delivered at Aeolian Hall in November 1917, Christabel argued:

> We have formed the Women's Party because our opinion has been, and is, that it would not be a good thing for women, the *new brooms in politics*, to go into those hopelessly dusty old places known as men's political parties (applause). What would our fight for the vote, and our dreams of what it would enable us to do, be worth if we were simply to go into the party political grooves which men have made, and which now the best of them are so anxious to get out of?[40]

Positing itself as the woman's voice in politics, the Party claimed that together women would cleanse the political sphere through their moral vision and courageous actions. Women's patriotic war work had provided one example of the ways in which women were uniquely fitted to preserve the nation; when the war ended they would use their votes to show the many other ways in which Britain's future depended on women. Shortly after the House of Lords voted in favour of women's suffrage in January 1918, while the war raged on, a *Britannia* editorial declared:

> When enemies of mankind are attempting to destroy our country by force from outside and by foul intrigue and demoralisation from within, it is the duty and the privilege of British women voters to set an example by their wise, self-disciplined, and constructive use of the vote.[41]

The Women's Party adopted the slogan 'Victory, National Security, and Progress' to signify its commitment to work for Britain's security and advancement. Many of the issues adopted and pushed most vehemently were matters not usually interpreted as women's particular concerns. But, within the minds of the Party leaders, international politics were not a purely masculine realm; women's interests were integrally bound up in the future of the British nation. The Women's Party called for more vigorous war measures, including the development of commercial kitchens, the reduction of non-essential industry, the elimination of all pacifist or pro-German civil servants, and better coordination of Allied military, naval, and aerial efforts.[42] The Party also advocated specific measures that would greatly expand women's power and influence, including equal employment opportunities for women, equal pay for equal work, and equal marriage and divorce laws. A central, though not fully articulated, feature was the ideal of a 'community', in which families would live together and share the tasks of cooking, cleaning, childcare, and education, thus reducing women's burdens.

The Women's Party platform, however, failed to attract enough votes in the November 1918 elections to send Christabel to Parliament. A strong endorsement from her wartime ally David Lloyd George and a record of support for the war effort helped bring her within 800 votes of Labour's candidate, J.E. Davison,[43] but ultimately she failed to appeal to the largely working-class constituency of Smethwick, an industrial suburb of Birmingham. Her radical suffragette reputation apparently still proved a hindrance in some minds. One journalist labelled the organization a 'Sex Party', predicated on their earlier 'sex war'. But Christabel retorted that the Party was 'not actuated by sex antagonism'. Their exchange points to a disagreement over more than just the meaning of the Party's name. The reporter's comment highlighted the common assumption that women, as they entered into full political participation in the nation, would simply be subsumed within its traditional, male-defined structures and ideals. Christabel, however, envisioned a far greater role for women, one premised upon her prophecy back in August 1914: 'a man-made civilization, hideous and cruel enough in time of peace, is to be destroyed. . . . Only by the help of women as citizens can the World be saved'. Now, with the vote in hand, women could assert their full authority and act upon an alternative, feminine vision for the nation.

The Women's Party, as such, challenged the longstanding belief that national strength and vigour depended only upon masculine ideals. In so doing, it exposed the socially constructed nature of women's traditional, passive relationship to the nation, as well as to war, and posited the possibility for change. However, Christabel, Emmeline, and their colleagues, in their work to redefine gender categories and widen women's influence, failed to foresee that, as they divested woman of her sex-differentiated functions and aligned her interests with those of the nation, they opened themselves to the charge of complicity in the destructive act of war.

Aftermath

Six years after the war, from the vantage point of her new home in California, Christabel surveyed Europe's troubled moral landscape. She witnessed a rising rate of illegitimate births, a rapidly changing sexual code among both men and women, and a massive questioning, brought on by the horrors of the war experience, of traditional political and religious authority. The British nation, as well, had lost some of its earlier lustre. The defeat of the German menace was a hollow victory, as Britain faced high unemployment, inflation, and the long process of rebuilding its population and infrastructure. As revolts and massacres spread through the Empire from Ireland to India, Britain no longer appeared to be an irresistible force of human progress. And women, despite their valiant war efforts, had not proved to be the future saviours of humankind. In the November 1918 elections, the more than eight million newly enfranchised women had failed to 'set an example' and had sent not a single politically viable woman to Parliament.[44] 'The shallow optimism of yesterday has vanished', Christabel lamented;

Various reasons forbid us to expect that, when other means are failing to save the world situation, the votes of women will succeed. Some of us hoped more from woman suffrage than is over going to be accomplished. My own large anticipations were based partly upon ignorance (which the late war dispelled) of the magnitude of the task which we women reformers so confidently wished to undertake when the vote should be ours.[45]

Women were not of one mind on world problems, she continued, and there was no distinctive women's point of view; 'And if there were, it would still be true of women, as of men, that they have not the wisdom to devise the right policy'.[46] Nor did they possess the right tools. Democracy appeared bankrupt and the vote 'going down in value, like Russian roubles or German marks. . . . The rifle of a black-shirted fascist, the industrial weapon of the striker, these are more potent in national affairs than the vote of a woman'.[47]

In those heady days of summer in 1914, suffragettes had firmly believed that women could and would bring about the 'harmonisation of the world'. Their confidence in women's moral superiority and spiritual link with the nation led them to assume that, once given the proper means, women could provide solutions to the problems 'created' by men. But the spectre of death and destruction during the war tempered such blind optimism, and demonstrated, ironically, that women were not essentially any different from men. They possessed no greater store of moral vision, no intrinsic link to the nation. Both sexes were trapped within an imperfect political system which would continue to allow international conflict, no matter how just one's own country. For all their virtues, women could work no miracles. As Christabel and many of her former suffrage colleagues went on to claim in the 1920s, that would have to be left to God.[48]

Notes

I would like to thank Sonya Michel, Billie Melman, Sybil Oldfield, and Anne Summers for their insightful comments on earlier drafts, as well as the attendees of the 'Women in Britain, 1914–1945' conference at Sussex University and participants in the 1991–92 Wiener Seminar at Tel Aviv University for their many provocative remarks.

1 *The Suffragette*, 7 August 1914, p. 301.
2 This idea was echoed through the writings of other WSPU suffragettes. See, for example, Mary Richardson's letter to M.A.R. Tuker, 10 January 1915, M.A.R. Tuker papers, Fawcett Library.
3 The National Union of Women's Suffrage Societies, under the leadership of Millicent Garrett Fawcett, temporarily suspended suffrage agitation and offered its services to war relief work; but this decision did not go uncontested and later resulted in a split in the executive committee. The United Suffragists was one of the few organizations to continue actively working for the vote. More information can be found in Johanna Alberti (1989), *Beyond Suffrage: Feminists in War and Peace, 1914–28*, New York, St Martin's Press; Sandra Holton (1986), *Feminism and Democracy: Women's Suffrage and Reform Politics, 1900–1918*,

Cambridge, Cambridge University Press; and Martin Pugh (1992), *Women and the Women's Movement in Britain, 1914–1959*, London, Macmillan.

4 As quoted in Andrew Rosen (1974), *Rise Up, Women! The Militant Campaign of the Women's Social and Political Unions, 1903–1914*, London, Routledge and Kegan Paul, p. 248 (emphasis added).

5 For more development of this theme, see Martha Vicinus, 'Male Space and Women's Bodies: The Suffragette Movement', in Martha Vicinus (1985), *Independent Women: Work and Community for Single Women, 1850–1920*, London, Virago.

6 Concern over national and racial degeneration was widespread in Great Britain, as it was elsewhere in Europe. See, for example, David Fraser Harris (1909), *National Degeneration*, Birmingham, Cornish Brothers. For a useful introduction to one aspect of this complicated issue, see Felicia Gordon, 'Reproductive Rights: The Early Twentieth-Century European Debate', *Gender and History*, 4:3 (Autumn 1992), pp. 387–99; also helpful is Richard Soloway (1982), *Birth Control and the Population Question in England, 1877–1930*, Chapel Hill, University of North Carolina Press.

7 The literature on pacifist women is vast. For an introduction, see Sybil Oldfield (1989), *Women Against the Iron Fist*, Oxford, Basil Blackwell; Ruth Roach Pierson (Ed.) (1987), *Women and Peace*, London, Croom Helm; Jo Vellacott, 'Feminist Consciousness and the First World War', *History Workshop Journal*, 23 (Spring 1987), pp. 81–101; Jo Vellacott, 'Anti-War Suffragists', *History*, 5:62 (October 1977); and Anne Wiltsher (1985), *Most Dangerous Women: Feminist Peace Campaigners of the Great War*, London, Pandora.

8 One important exception is Anne Summers (1988), *Angels and Citizens: British Women as Military Nurses, 1854–1914*, London, Routledge and Kegan Paul; see also Anne Summers, 'Militarism in Britain before the Great War', *History Workshop*, 2 (Autumn 1976), pp. 104–23.

9 The most salient example here is Andrew Rosen's *Rise Up, Women!* (see note 4).

10 See Jo Vellacott, 'Feminist Consciousness and the First World War', and the following discussion in Ruth Roach Pierson (Ed.) (1987), *Women and Peace*, London, Croom Helm.

11 *Ibid*. See also Claire Tylee (1990), *The Great War and Women's Consciousness: Images of Militarism and Womanhood in Women's Writings, 1914–64*, Iowa City, University of Iowa Press, p. 37; Martin Pugh attempts to correct this tendency in *Women and the Women's Movement in Britain, 1914–1959* (London, Macmillan, 1992), pp. 10–11.

12 See, for example, Jean Bethke Elshtain (1987), *Women and War*, New York, Basic Books; and Sandra Gilbert and Susan Gubar's essays in Margaret Higonnet *et al.* (Eds) (1987), *Behind the Lines*, New Haven, Yale University Press.

13 Antoinette Burton, 'The Feminist Quest for Identity: British Imperial Suffragism and "Global Sisterhood", 1900–1915', *Journal of Women's History*, 3:2 (Fall 1991), p. 46.

14 *Ibid.*, p. 69.

15 In 1915 Christabel's speech was published in pamphlet form by the Women's Press under the title 'America and the War'.

16 *Ibid.*, p. 15.

17 *Ibid.*

18 For further development of these issues, see 'Mrs Besant on the War', *The Suffragette*, 18 June 1915, p. 149.

19 See for example, Philippa Levine (1990), *Victorian Feminists*, Oxford, Basil Blackwell, p. 27; L. Whitelaw (1990), *The Life and Rebellious Times of Cicely Hamilton*, London, The Women's Press, p. 56; Andro Linklater (1980), *An Unhusbanded Life*, London, Hutchinson, pp. 53, 100 (my thanks to Anne Summers for bring the latter two references to my attention). In the 1860s and 1870s, when

first articulating their feminist views, many British women found inspiration in Italian political theory. Numerous references to Mazzini are sprinkled throughout *The Englishwoman* during those years and continue to appear in *The Suffragette* into the twentieth century.

20 Christabel's use of this idea to substantiate her cause against the Germans is ironic: in 1801 Frederick William II of Prussia originally articulated the idea that love of fatherland is a preparation for love of mankind. See George Mosse (1985), *Nationalism and Sexuality: Middle-Class Morality and Sexual Norms in Modern Europe*, Madison, University of Wisconsin, p. 71.

21 C. Pankhurst, 'America and the War'.

22 The last accurate financial report was issued in the spring of 1914 for the fiscal years ending 28 February 1914, so it is difficult to provide figures of the intake or expenditure of WSPU funds (Wiltsher, *Most Dangerous Women*, pp. 38–9; Rosen, *Rise Up, Women!*, p. 252.)

23 In December 1915 the SWSPU published the first issue of the *Suffragette News Sheet*, which appeared monthly from April 1916 to November 1918.

24 The IWSPU had branches in Chelsea, Glasgow, Hammersmith, Hampstead, Preston, and Richmond in Surrey, although membership figures for each are unavailable. Charlotte Marsh was the organizing secretary.

25 *The Suffragette*, 16 April 1915, p. 3.

26 *Ibid.*

27 *The Suffragette*, 9 July 1915, p. 198.

28 *Ibid.*

29 *The Suffragette*, 9 July 1915, p. 198.

30 As reported in *The Suffragette*, 25 June 1915, p. 165.

31 See, for example, *The Suffragette*, 21 May 1915, p. 92. For a broad and convincing analysis of the relationship between nationalism and Protestant, middle-class respectability, see George Mosse, *Nationalism and Sexuality*.

32 Mosse, *Nationalism and Sexuality*, p. 98.

33 'What a German Victory Would Mean', *The Suffragette*, 27 August 1915, p. 290.

34 For examples of women's feminizing influence in countries under siege, see Mrs. St. Clair Stobart, *The Flaming Sword in Serbia and Elsewhere*, 2d. ed., (London, 1917), and May Sinclair, *A Journal of Impressions in Belgium* (London, 1915).

35 'The Women of the Enemy Country Are Our Enemies', *The Suffragette*, 16 April 1915, p. 4; 23 April 1915, p. 25.

36 *Britannia*, 5 November 1915, p. 41.

37 Mrs Pankhurst had met Madame Botchkareva while travelling in Russia with Jessie Kenney during the summer of 1917 on a mission to 'explain to the Russian people the opinions as to the war and the conditions of peace held by us patriotic British women' (*Britannia*, 6 June 1917, p. 3). With the good will and perhaps even the financial backing of David Lloyd George, the two left in early June and returned in September 1917. While their conferences with numerous Russian leaders, including Aleksander Kerensky, were disappointing, their contact with Madame Botchkareva proved inspiring. More can be found on Madame Botchkareva in Isaac Don Levine (1919), *Yashka: My Life as Peasant, Exile and Soldier*, London, Constable; and in Wiltsher, *Most Dangerous Women*.

38 In January 1918 the House of Lords followed the Commons' lead, and the Representation of the Peoples Act became law on 6 February 1918, when it received royal assent.

39 Mrs Flora Drummond became the Chief Organizer, Mrs Pankhurst became treasurer, Annie Kenney became secretary, and Christabel continued to hold her powerful position as editor of *Britannia*, which remained the party's organ.

40 *Brittania*, 30 November 1917, p. 207 (emphasis added).

41 *Britannia*, 18 January 1918, p. 249.

42 *Britannia*, 2 November 1917, p. 171.
43 Davison won 9,389 votes while Christabel attracted 8,614, a margin of only 775. (Rosen, *Rise Up, Women!*, p. 269.)
44 The one woman elected, Countess Markiewicz, refused as a member of Sinn Fein to take her place in Parliament.
45 Christabel Pankhurst (1924), *Pressing Problems of the Closing Age*, London, Morgan and Scott, p. 38.
46 *Ibid.*, p. 42.
47 *Ibid.*, p. 39.
48 As David Mitchell has noted in *Queen Christabel* (London, Macdonald and Jane's, 1977), many of the most prominent WSPU women, including Christabel Pankhurst, Annie Kenney, and Flora Drummond, turned to religion during the 1920s.

Chapter 8

England's Cassandras in World War One

Sybil Oldfield

In February 1915 Aletta Jacobs, Holland's first woman doctor, pioneer of birth control and leading suffragist, cabled women's organizations all over the world calling for an International Women's Congress to protest against the 'Great War' and to try to prevent any recurrence: 'We feel strongly that at a time when there is so much hatred among nations, we, women, must show that we can retain our solidarity'. Four months later, in June 1915, a list was published in London[1] with the names of 156 British women who had supported Dr Jacobs in holding that Women's International Congress at The Hague at the end of April.[2] The Congress had set out two aims:

1   To demand that international disputes shall in future be settled by some other means than war.
2   To claim that women should have a voice in the affairs of the nations.

At first I assumed that those 156 names represented all the British women who had wished to attend the Congress at The Hague; however, a fellow researcher, Hilary Frances, of Harrogate, Yorkshire, then discovered another annotated list, consisting of 180 names, among the Catherine Marshall Papers in Cumbria Public Records Office, Carlisle, and headed in Catherine Marshall's handwriting: '*Private*: *for Mr McKenna*' (the Home Secretary), *c.* 17 April 1915. It was this second list that in fact gave the names of all those women who had applied for Exit Permits to cross the Channel in wartime in order to attend the International Congress in neutral Holland. My previous list, I now realized, consisted of those British women willing to come out in public in support – including financial support – of this anti-militarist initiative. Some women were on both lists, others only on one. (It became clear, on analysis, that poorer women delegates could not afford to be financial sponsors, while some of the latter were too old, or too committed to the care of sick relatives or children to be free to attend in person.) Collating the two lists yielded 283 names. Subtracting official reporters and observers left 275. Of these, some were very well-known: for instance, Mrs Despard, Eva Gore-Booth, Emily Hobhouse, Lady Ottoline Morrell, Mrs Pethick-Lawrence, Sylvia Pankhurst, Maude Royden, and Olive Schreiner. Others would have been well known in their own organization or locality but have since been forgotten; still others were obscure then as now. Believing that to dissent in public from Britain's war-fever in 1915 was the hardest and most vital act of moral courage that any citizen could then undertake, and that it was particularly difficult for women to seem ungrateful towards the men then volunteering for mutilation and death,[3] I have attempted to analyze all the names on the two lists. So far I believe that I have correctly identified 200 of the 275.

Several questions had to be addressed concerning the composition of this dissident group and their political ideologies. A study of their class background suggested that 50 per cent came from the professional, educated middle class, 27 per cent from the wealthy upper middle class, 15 per cent from the gentry and aristocracy, 5 per cent from the working class and 3 per cent from the lower middle class. Such a solid preponderance of educated and/or professional women is not surprising. What is astonishing is the high proportion, relative to the general population, of very wealthy and/or upper-class women who crossed the traditional political class-divide at this time of national crisis and identified themselves as dissidents. One explanation emerged when the women's religious background was uncovered, since many came from a network of rich Quaker families. That disproportionately few women came from lower-middle-class or working-class backgrounds is *not* evidence for a lack of support among poorer women in Britain for moves to end the first world war as soon as possible by negotiation – one has only to remember the Women's Peace Crusade led from Glasgow, 1916–1918[4] and Ada Nield Chew,[5] Selina Cooper,[6] Hannah Mitchell,[7] and Alice Wheeldon.[8] But those women had almost no money and this particular transnational idealistic initiative, involving huge travel charges and organizational expenses, could only be backed by women with access to some disposable income of their own – with the exception, that is, of the few working-class delegates from large organizations such as the Women's Co-operative Guild or the Women's Labour League. It would, therefore, be a serious distortion of the truth to claim that the backers of the Women's International Congress at The Hague *were* the Women's Peace Movement in Britain in World War One. What they did constitute was simply its most socially prominent leadership and prophetic voice.

Regarding their work, either paid or unpaid, 10 per cent of the women were fully occupied in the home as young mothers or elderly wives and grandmothers, thirty-two of the two hundred women being over 60. At least 50 per cent worked voluntarily outside the home in various branches of social reform such as the improvement of inner-city housing conditions, adult literacy classes, the liberalization of penal policy towards young offenders, initiatives relating to alcoholism, the organization of Youth Clubs for factory girls, or the establishment of refuges for homeless women (*plus ça change*). Ten per cent of the women held elected unpaid positions as Poor Law Guardians, local councillors, political workers or trades unionists – for example, Councillors Margaret Ashton, Susan Lawrence, Eva McLaren, and Ada Salter; the Poor Law Guardians Henrietta Barnett and Sarah Reddish; the Hon. Secretary of the National Union of Women's Suffrage Societies, Kathleen Courtney, and the Hon. Secretary of the Women's Co-operative Guild, Margaret Llewelyn Davies. Ten per cent were health workers, including the midwife Edith Pye, the nurse Violet Tillard, and five doctors – Elizabeth

Knight, Ettie Sayer, Barbara Tchaykovsky, Henrietta Thomas, and Ethel Williams. Finally, 20 per cent could be loosely categorized as intellectuals, including many teachers and the outstanding classicists Melian Stawell and Louise Matthaei, the mathematicians Julia Bell and Frances Hardcastle, the expert on international labour law Sophy Sanger, the historian Alice Clark, the theologian Olive Wyon, the artists Beatrice Collins, M. Sargent, Florence and Janet Robertson, the musicians Winifred Holiday and Marjorie Kennedy Fraser, and, finally, such writers and journalists as Evelyn Sharp, Molly Hamilton, Caroline Playne, and Irene Cooper Willis.

Political analysis of the 200 women is rather more complex. Since they were all anti-militarist feminists and on the liberal/left side of the party political divide it might seem sufficient simply to term them all 'radical suffragists' or 'pacifist feminists'. But they do not in fact fit into any such unitary category. Some of the women, for example Sylvia Pankhurst, Maud Joachim, Alison Heilans, Dorothy Evans, and Muriel Matters, had been militant suffragettes, not constitutional suffragists, before 1914. And far from all of the women were absolute pacifists. Even more important, although they were all radicals and feminists, by 'coming out' as anti-militarists in the middle of World War One, they had to break away not only from the great majority of their fellow radicals in the Liberal and Labour Parties but also from the great mass of the Women's Movement. 'Women, your country needs you', said Mrs Fawcett – and most women in Britain went along with her appeal to their life-sustaining patriotism on the home front. When the pacifist women in leadership positions in the National Union of Women's Suffrage Societies, Catherine Marshall, Kathleen Courtney, Maude Royden, Helena Swanwick, and Crystal Macmillan, among others on The Hague Congress list, resigned in protest against Mrs Fawcett's taking the Women's Suffrage Union into the national war effort, they were all subsequently defeated in their attempt to get re-elected onto the NUWSS's Executive on an internationalist, anti-militarist platform. Clearly these suffragists wore their feminism with a difference by 1915.

If one asks how these 200 women would have identified themselves ideologically *before* August 1914, in my view 45 per cent would have declared themselves to be feminists first and foremost, and another 10 per cent would have called themselves socialists. Possibly as many as 25 per cent would have said that they were practising Quakers for whom faith in the Inner Light within all people was the central tenet of their lives. And the last 20 per cent would have had to be labelled 'idealistic humanitarians' – women with a compulsion to intervene in any and every case of remediable suffering – not just women's suffering – that crossed their path – for example, Emily Hobhouse, the campaigner against British concentration camp policy in the Boer War, or the Christian Socialists Henrietta Barnett, Dorothy Buxton, and Mary Hughes. What then happened in the months immediately after August 1914 was that the 55 per cent who had hitherto been feminists or socialists first, went over to join forces with the other 45 per cent of Quakers and humanitarians – even at the cost of breaking with their own former friends and comrades. They were convinced that the need to end World War One – and in such a way that it would not cause a World War Two – was *the* supreme issue for our whole species. As the feminist Helena Swanwick wrote in retrospect, 'If I had felt driven to fling myself into the movement for the vote, I was even more ruthlessly compelled to discover the truth, as I saw it, about war in general and this war in particular'.[9]

Helena Swanwick and all her fellow anti-militarists who supported the Women's International Congress at The Hague were, of course, pilloried in the popular press as 'pro-Hun peacettes', 'feminine busybodies', 'hysterical women', 'babblers, folly in petticoats', 'peace fanatics', 'mischievous, futile, blundering Englishwomen', etc., etc. But they were also extremely unpopular among their previous allies in the campaign for the vote. As the suffragist Wilma Meikle wrote in 1916:

> if over here the pacifists had had their way and had succeeded in identifying the various suffrage societies with their own cause, the result would certainly have been the total and lasting collapse of the suffrage movement.
>
> Fortunately, that identification was never effected. The faint voices of the pacifists had cried in a wilderness and 1915 found thousands of women offering themselves for war-work.[10]

Among those 'criers in the wilderness' were a few women, all supporters of the Congress at The Hague called by Dr Aletta Jacobs, who felt compelled to speak out like Troy's Cassandra, prophesying to deaf, mocking ears that the governments of the Great Powers were leading the world into chaos and perhaps ultimate extinction. The first such note of horrified foreboding occurs in the unpublished journals of Beatrice Webb's forgotten sister, Kate Courtney, the wife of Leonard Courtney, the veteran pacifist Liberal peer. Already in July 1911, during the Agadir crisis, Kate Courtney had realized that Italy's seizure of Tripoli from the Turks would 'put back the Peace ... movement terribly and set a precedent which will greatly strengthen mutual suspicion and add to armaments'. She recognized the danger inherent in all the guilty imperialisms, with Germany growing more and more bitter as 'a poor lion without a Christian'. By 1912 Kate Courtney noted in her diary all three factors that would soon contribute to catastrophe – secret cabinet government which could mobilize the country without warning; secret international diplomacy based on fears and animosities; and the uncontrolled armaments race between the rival Powers. On 1 August 1912 she wrote:

> One or two bad [Parliamentary] debates on Navy [i.e. ordering more Dreadnought battleships]. It is almost enough to make one despair – the Government's poverty of resource or effort towards a real understanding with Germany. It seems as if we were assisting at a Greek tragedy and some terrible catastrophe was nearing us every day – and for no reason – just blind fate – insanity.[11]

Very soon after World War One broke out, a second Cassandra emerged: Mary Sheepshanks, the editor of the International Women's Suffrage Movement periodical *Ius Suffragii*. In a signed editorial headed 'Patriotism or Internationalism', Mary Sheepshanks denounced the irrational killing-competition on all sides: 'Each nation is convinced that it is fighting in self-defence, and each in self-defence hastens to self-destruction'. Already she saw that 'the world is relapsing into a worse, because a more scientific barbarism than that from which it sprang'. She concluded her article in November 1914 with the urgent warning:

> Armaments must be drastically reduced and abolished, and their place taken by an international police force. Instead of two great Alliances pitted against each other, we must have a true Concert of Europe. Peace must be on generous, unvindictive lines, satisfying legitimate national needs, and leaving no case for resentment such as to lead to another war. Only so can it be permanent.[12]

A third Cassandra who supported the Congress at The Hague was the art historian Vernon Lee, of whom Irene Cooper Willis wrote:

> It was rare, in the pre-war period, to find a writer and an aesthete so in touch with European liberal opinion ... and so alive to the various national policies which led to the Great War. ... She felt the war deeply, and was torn by it more than most people, because she had roots in Germany, as well as in England, Italy and France.[13]

All her attempts to warn liberal opinion of the pitfalls of nationalism in her letters and journalism in all her four countries[14] having failed, Vernon Lee was indefatigable in her entreaties that the war be ended immediately and in such a way as not to be the inevitable cause of a sequel. On 17 September 1914 Vernon Lee published a message to Americans in the New York *Nation*, in which she countered H.G. Wells' plea to Americans in the *Daily Chronicle* of 24 August to 'STOP VICTUALLING OUR ENEMIES'. Wells was asking neutral America to lend England 'the weapon called *famine* and famine-sprung disease', wrote Vernon Lee. And if America were to help to end the war by destroying Germany, all that would be achieved would be the reinforcement of Prussian militarism and absolutism and its drive for revenge. On 3 October 1914 the New York *Evening Post* published Vernon Lee's open letter to Rosika Schwimmer – who was to be one of the driving forces behind the Women's International Congress at The Hague the following April. In that letter Vernon Lee supported Rosika Schwimmer's call for continuous mediation between neutrals 'to stop the international massacre at the earliest possible moment'. What Vernon Lee constantly insisted on was the necessity for 'an endurable, and therefore enduring peace'. In her open letter, Vernon Lee complained that she had been refused the opportunity to try to warn her compatriots against their deluding myth of total self-righteousness and the total war guilt of the other side, 'by the once liberal and radical ... papers of my country. ... Similar self-justificatory myths ... have doubtless arisen in every one of the belligerent countries'. Only the ILP and its paper, the *Labour Leader*, were willing to let her speak. 'We in England have no chance of hearing the truth except from the lips of neutral ... nations. ... This war has strangled truth, and paralyzed the power and wish to face it'. Already in October 1914, less than eight weeks into the Great War, Vernon Lee was prophesying: 'It is not the diplomatists and soldiers who can end ... this butchery and destruction ... otherwise than in some new sowing of dragon's teeth'. Therefore she begged:

> It is the peaceful interests which have been sacrificed, the human affections which have been violated; it is the network of international cooperation in trade, in art, in science, and in progress which has been rent and trampled in blood and mud; it is those who need peace and believe in peace through peace who alone should end this war.

On 1 April 1915 Vernon Lee published her internationalist-humanist credo in the *Labour Leader*, warning her fellow citizens against the folly of regarding the Germans as anything other than our German selves:

> [What] we are treating merely as a nuisance to us, is a creature as like ourself as our own image in the glass; ... strike out at it, and it strikes back at you ... I note with satisfaction that Miss Christabel Pankhurst and her militants with their idiotic destructive heroism are up in arms against German militarism. Their spirit is now in the whole nation – the spirit which overlooks the fact that your adversary is human like yourself, and will not yield to methods you yourself would never yield to.

Already in 1915 Vernon Lee anticipated an Allied victory that would impose a crushingly punitive settlement on Germany. Therefore she felt driven to publish what she knew would be a very unpalatable prophecy of doom. She called her pamphlet *Peace with Honour – Controversial Notes on the Settlement*. Her publisher, the Union for Democratic Control of Foreign Policy, prefaced it with the disclaimer that it did 'not necessarily adopt as its own every statement or opinion therein contained'. Among her unpopular propositions were:

1 The decision by arms proves nothing except that the victor is the victor and the vanquished the vanquished.
2 Therefore in the 'erroneously called ... arbitrament of war, [the victor] unites the position of judge, jury, policeman and plaintiff' in demanding compensation from the vanquished.
3 'War compensations and penalties [including imputation of war-guilt] are really nothing but advantages extorted by the victor.'
4 A huge war-indemnity imposed upon a defeated Germany would be both self-defeating in its damage to her trading partners and, 'far from "crushing" German militarism, England and her Allies [would have] hit upon an infallible recipe for giving it a new lease of life. A humiliated, insecure, or hemmed-in Germany would ... mean a Germany arming once more for a Leipzig after a Jena'.

How could Vernon Lee have been so certain of her ground in thus warning her compatriots so precisely against the as yet unwritten stipulations of the Treaty of Versailles? She anticipated the question by insisting in her pamphlet that

The question is a mainly psychological one, it is one of probable feelings, desires and effects. And psychology is merely the study of human nature by means of observation of our own thoughts and feelings.... We have therefore [only] to ask ourselves 'How should we feel and behave if a victor ... tried to crush us?'

In likening us to 'the enemy' and them to us, Vernon Lee was acting upon her *a priori* assumption that all humans are profoundly alike, and never more tragically alike than when postulating the existence of a devilish 'Other'. In fact, Satan the Waster or War-Bringer lives inside our own heads and is happy to recruit even our most tender feelings of pity and indignation whenever necessary to validate righteous hatred and our own mass killings.[15]

In 1917 a new voice, that of the Quaker idealist Marian Ellis,[16] expressed her prophetic fears concerning the post-war decisions soon to be made by victors and losers alike. Starting from the principle that every human being on earth makes a unique contribution to the world and that therefore human justice is incompatible with the institutionalized killing of humans, Marian Ellis reminded her readers:

History is full of instances of an ideal being lost through the methods of its advocates. We are always trying to cast out Satan, if not by Satan exactly, at least by one of the smallest of his angels.... Nations are but communities of men and women.... Can we make an effective appeal to [the] Inner Voice in the heart of the other man, in the hearts of the people of the other nation, if our [own] hands are full of the instruments of destruction, even in the supposed interests of justice?[17]

To Marian Ellis, disarmament was not a millennial goal but rather the immediate next step – if humanity is to be saved from its misguided rulers:

At the end of this war the world will have to decide which way it desires to go, towards disarmament *or* destruction; there will be no middle course. The choice is really between basing our civilization on faith or on fear, and the question of armaments stands at the parting of the ways.... The world will be weary to death of war when this war is over.... But however great the reaction against war may be, it will not suffice to bring about the state of mind which produces disarmament, unless it is rooted in a positive faith that can overcome fear.... [For] Armaments are the visible and tangible sign of our lack of faith in ... our fellow men.... The armies and navies stand now between the opposing nations like a great barrier preventing the vital realization of the unity of the spirit.... Disarmament is not merely scrapping our guns and our battleships. It is the working out of a national policy which, being inspired by love for all men, cannot be antagonistic, ... it is the problem of India, of Ireland, of our relations with Russia and Persia, Germany and Belgium, as God would have them to be.[18]

Marian Ellis

Addressing herself to what was being perpetrated by Britain in 1917, Marian Ellis denounced the anti-Christian cruelty and humbug of her own side: 'We pray that we may love our enemies, and meanwhile we are involved in the crime of starving [German] women and children. We pray God to turn their hearts, while our men slay their bodies'.[19] She aligned herself with the absolutist Quaker stand on pacifism as articulated by the Meeting for Sufferings in London, January 1917, as it warned, in words that she helped to draft, 'we must either stand for an unweaponed faith, the abolition of armies and navies, and reliance upon spiritual forces alone, or resign ourselves to a still more complete organization of the world for war'.

What it felt like to have been an unheeded, derided Cassandra during World War One has been best described by another supporter of the Women's International Congress at The Hague, Helena Swanwick. On 4 August 1914 she attended a women's protest meeting that called on British women to 'down tools' and so effect a general strike against the sudden war. But Helena Swanwick knew all such calls to be futile – women would never strike against their own helpless dependants, their children, their sick or their elderly. In her memoir, *I Have Been Young*, of 1935, Helena Swanwick recalled her own compulsion to resist the engulfing war-fever and her simultaneous certainty that all such resistance was useless:

> Fighting in me on that evening and for four years to come was, on the one hand the conviction that all we [women protesters] said and did would be treated by the mass of our fellow-countrymen at best as 'twittering' (Mr Asquith's word), at worst as treachery, and on the other hand, the conviction that 'I could do no other.' I was 'driven', as I had been in 1906 [to speak out for women's suffrage]. But whereas then there had been unquenchable hope and buoyant comradeship, there was now a rending pity, a horror of black darkness, and in my brain, almost physically audible at times and never ceasing, something like a monotonous bell for ever tolling: 'Wicked! Wicked! Wickedly silly! Cruel! Silly! Silly!' ... But I was quite sure (and said so on countless platforms and wrote it in many articles) that the [so-called] 'knock-out' [blow] would end in 'an inconclusive peace'. I was quite certain that a prolonged war, ending in a decisive victory for one side or the other, would result in brutal demands for 'Annexations and Indemnities', and I trusted our politicians, and especially Mr Lloyd George, to disguise them under the suaver names of 'Mandates', 'Self-determination', 'Reparations' ... I spoke a great deal in public during the four years that followed, but I never said a hundredth part of what I thought and felt. Not because I was afraid of abuse or violence – I experienced plenty of both – but because I was afraid of failing altogether to be comprehensible. It was lonely in those days ... We failed. We could not overtake the lies; the disastrous knock-out blow had the anticipated consequences from which the world will suffer for a century or more.

Even after the punitive clauses in the Peace Treaty of 1919 were published, these Cassandras and their fellow anti-militarists in Britain and on the

Continent attempted to make one last concerted attempt to avert a second 'Great War' in Europe. They held another Women's International Congress, this time in Zurich, in 1919, at which they were the first people in the world to identify the fatal weaknesses in the proposed League of Nations. The non-admittance of Germany and the Soviet Union at the start, the non-implementation of the promised general disarmament clauses, and the unworkable sanctions provisions – each of these factors leading to the failure of the League was spelt out in advance by the women in Zurich. They cabled President Woodrow Wilson and the other leaders at Versailles, warning them that

> 'The terms of peace can only lead to future wars. . . . By the financial and economic proposals a 100 million people of this generation in the heart of Europe are condemned to poverty, disease, and despair, which must result in the spread of hatred and anarchy within each nation.

Kate Courtney prophesied on 9 January 1920 in a letter to *The Daily Mirror*:

> The cruel and unwise conditions of the Peace Treaty insisted on by France will have results [which] will be suicidal to France . . . and disastrous to all Europe . . . M. Clémenceau is a great Frenchman, as Bismarck was a great German. Both in their hour of victory have done an evil thing for their respective countries and for the world.

Simultaneously, the Quaker Joan Fry, working to organize the feeding of German civilians still suffering from the effects of the Allies' economic blockade, was prophesying that the Allies were creating more and more reactionaries in Germany. And just three years later she reported to the Society of Friends: '150 million Deutschmarks for a quarter of a pound of butter. Martial law in Berlin. In Nürnberg I saw seven men in the new Hitler uniform'.[20]

Unlike the original Cassandra of Greek mythology, these English Cassandras were not annihilated by the horrors outside themselves or their despair within. Each time that they failed to win a hearing – and they were under no illusion but that they did fail – they summoned up the energy for new effort, both mental and physical. Almost from the first day of the war Kate Courtney worked for the relief of ordinary German civilians, trapped in England as destitute 'enemy aliens' between 1914 and 1918, founding an Emergency Committee for War Victims; and at the end of the war it was in her house in Chelsea that Kate Courtney, then aged 71 and recently widowed, helped to found the Fight the Famine Committee, out of which grew the Save the Children Fund.[21] Both Vernon Lee and Helena Swanwick worked for the radical 'think-tank', the Union for Democratic Control of Foreign Policy, discussed and affirmed by A.J.P. Taylor in *Troublesome People*; Mary Sheepshanks worked for the resettlement of hundreds of thousands of Belgian refugees in Britain for the war's duration and she also continued to edit her International Women's Movement monthly paper, *Ius Suffragii*, distributing it across enemy frontiers via neutrals throughout the war; Marian Ellis gave away her personal fortune to support the imprisoned 'absolutionist' conscientious objectors and their impoverished families via the No Conscription Fellowship and the Quaker Friends' Service Committee.[22] Joan Fry became a Quaker Minister to imprisoned Conscientious Objectors, monitoring their conditions and, in certain cases of gross maltreatment, effectively saving some men's lives by procuring their release just in time. Immediately after the war Joan Fry began her immense four-year-long relief project to feed starving German children and students, which became known colloquially in Germany as 'Quakern'.[23]

There are many different stories to be told about the 200 women, so far identified, who answered Dr Aletta Jacobs' Call to the Women of the World in February 1915. I have singled out just six who seem to me to have been doomed to have prophesied truly but in vain. The rest of the twentieth century has indeed enacted that Greek tragedy that Kate Courtney saw approaching in 1911. The world has 'relaps[ed] into a worse, because a more scientific barbarism than that from which it sprang' as Mary Sheepshanks foretold. Germany was not crushed by the punitive 'Peace' but rather won over to a renewed militarism – 'a Germany arming once more for a Leipzig after a Jena' as Vernon Lee, Helena Swanwick, and Joan Fry all warned. The ideals, both of liberals in the West and communists in the East, have indeed been lost through the methods each employed and the alternative to disarmament is still destruction – truths even more glaring to us at the end of the century than they were to Marian Ellis in 1917. The positive injunctions of these women may be as worth heeding as their rejected warnings. We do need a *dis*armament race worldwide; we do need an effective, impartial, and therefore respected international UN policing force. Perhaps we should even give aid, not withhold it, from the subjects of dictatorship, acting on the basic principle of fellow-humanity, instead of punishing the victims for the crimes of their rulers.

Doubtless they were not saints, these Cassandras, neither perfectly wise nor perfectly good. And yet their words, buried in old newspapers, in unpublished journals and out-of-print books, come across as extraordinarily alive and nourishing. They seem to confirm Simone Weil's declaration during the Second Great World War that 'We can find something better than ourselves in the . . . past'.[24]

Notes

1 The list of names was published on the back page of *Towards Permanent Peace: A Record of the Women's International Congress*, June 1915, London.

2 In fact only three Englishwomen managed to reach The Hague because the Admiralty closed the North Sea to British shipping. Nevertheless, a remarkable Congress did take place consisting of 1,300 women from twelve countries, presided over by the American, Jane Addams. At its close, the Congress despatched women envoys to all the Foreign Ministries of Europe's belligerent and neutral nations, asking the belligerents to state their war aims and the neutrals to attempt continuous mediation – of course in vain. See A. Wiltsher (1985), *Most Dangerous Women*, London, Pandora.

3 Cf. 'Even to seem to differ from those she loves in the hour of their affliction has ever been the supreme test of a woman's conscience': Jane Addams, in J. Addams,

 E. Balch, and A. Hamilton, (1915), *Women at The Hague*, New York, Macmillan, p. 125.

4 See J. Liddington (1989), *The Long Road to Greenham*, London, Virago, pp. 107–30.

5 See D. Nield Chew (1982), *The Life and Writings of Ada Nield Chew*, London, Virago.

6 Liddington J. (1984), *The Life and Times of a Respectable Rebel: Selina Cooper 1864–1946*, London, Virago.

7 H. Mitchell (1968), *The Hard Way Up*, London, Faber.

8 S. Rowbotham (1986), *Friends of Alice Wheeldon*, London, Pluto Press.

9 H. Swanwick (1935), *I Have Been Young*, London, Gollancz.

10 W. Meikle (1916), *Towards a Sane Feminism*, London, Grant Richards, pp. 154–5.

11 Kate Courtney's diaries are held among the Courtney Papers, British Library of Economic and Political Science.

12 See S. Oldfield (1984), *Spinsters of this Parish: The Life and Times of F.M. Mayor and Mary Sheepshanks*, London, Virago, ch. 9.

13 Preface to *Vernon Lee's Letters* (1937), privately printed.

14 P. Gunn (1964), *Vernon Lee*, London, Oxford University Press, ch. 14.

15 V. Lee (1920), *Satan the Waster*, London, John Lane, praised by Shaw in *The Nation*, September 1920.

16 Marian Ellis (1878–1952) was one of the heirs to the Rowntree estate and daughter of the Liberal MP John Ellis. After the Boer War she helped Ruth Fry in relief work among South African women; after World War One she became Secretary of the Fight the Famine Council whose Chairman, Lord Parmoor, she married, thus becoming the stepmother of Stafford Cripps. During the 1920s Marian Parmoor became President of the British National Peace Council and the British Section of the Women's International League for Peace and Freedom. After World War Two she studied nuclear fission in order to speak with some authority concerning the abuse of atomic energy. Two days before her death in 1952 she was helping to draft a Quaker message to the Prime Minister, protesting against the Allied bombing of North Korea.

17 'The Spiritual Aspect of International Unity. A Plea for the Principle of Consent', 1917. Friends' House Library, London.

18 Marian E. Ellis, 'Disarmament. (i)', *Friends' Quarterly Examiner*, 1917, pp. 182–7.

19 *Ibid.*

20 Unpublished letter held in the Joan Fry Archives, Friends' House, London.

21 See S. Oldfield (1989), *Women Against the Iron Fist: Alternatives to Militarism 1900–1989*, Oxford, Basil Blackwell, ch. 2.

22 See T. Kennedy (1981), *The Hound of Conscience: A History of the No-Conscription Fellowship*, Fayetteville, University of Arkansas Press.

23 See Joan Fry Archive, Friends' House Library, London.

24 S. Weil (1962), *Selected Essays 1934–1943*, Oxford University Press, p. 44.

Chapter 9

Women in the British Union of Fascists, 1932–1940

Martin Durham

While Fascism in Britain dates back to the early 1920s, its most important expression in the inter-war period, the British Union of Fascists, first saw the light of day in October 1932. Led by the former Labour Minister, Sir Oswald Mosley, by the middle of 1934 the BUF had gained in the region of 40,000 members. But the movement's growth was to be short-lived and the violence at a meeting in Olympia addressed by Mosley led to widespread denunciation and a haemorrhage of members which pushed the BUF back to the political margins. Despite decline, however, the movement continued to be active and, relying heavily on anti-semitism, it succeeded in building up significant support in the East End of London. Changing its name first to the British Union of Fascists and National Socialists, then to British Union, it vehemently opposed war with Nazi Germany and, towards the end of the 1930s, apparently enjoyed a revival. But the coming of war was to prove fatal to the movement, and in mid 1940, viewed as a security risk, the organization was proscribed by government order and many of its activists interned.

The BUF has been the subject of considerable attention by historians. Much has been written about both the leading figures of the organization and the rank and file Blackshirts who sold papers, spoke on street corners, and went on marches. Yet in one respect we have barely begun to get to grips with the British Fascist movement of the 1930s. Numerous writers have emphasized the patriarchal character of the extreme right, the links between *fascismo* and *machismo,* and just as Goebbels had characterized National Socialism as essentially 'a masculine movement' (Millett, 1971, p. 165) so too we can find such a view articulated within the British Fascist movement. Thus, for one leading Mosleyite, A.K. Chesterton, Fascism represented a 'Return to Manhood', the re-assertion of 'the masculine spirit' against 'the matriarchal principle' (*Action*, 9 July 1936). But this was not the only face of the movement. The photographs and newsreels of the time, as we might expect, show massed arrays of male Blackshirts. But, as the visual evidence makes clear, it was not only men who donned the black shirt and, if we examine the BUF press (*Fascist Week, Blackshirt, Action*), we find an abundance of material on how the movement sought to appeal to women and the activities of those it gained. In this chapter, I shall explore a neglected aspect of women's political involvement, and also question the existing assumptions of the relationship between women and Fascism.

In March 1933, six months after its inception, the British Union of Fascists established a Women's Section and in an article published the following year, its founder, Lady Makgill, noted how starting with only 'a few dozen women, scattered about the country', the organization had developed to such a state that it now had its own Women's Headquarters building and 'women's branches in seventy-five per cent of the districts where the men's branches are established' (*Age of Plenty*, vol. II, no. 1 (1934)). The Blackshirt press carried numerous accounts of these branches and their activities. Worthing, for instance, had its own quarters, as did the Brighton and Hove branch (*Blackshirt*, 8 June 1934; 7 December 1934) and it would be here or on other premises that women members would be educated in the Union's policy and in some cases trained as public speakers. According to an account in mid 1934, 'every branch' held either a Study Circle or a Speakers' Class, and some both. Streatham, we are told later in the year, held both, while Ilford had begun a weekly policy class in the home of one of its members (*Blackshirt*, 1 June 1934; 28 September 1934; 12 October 1934). At Women's Headquarters at the beginning of the following year both were available: Policy Class twice on Friday, while on Wednesday the public speaking class covered 'correct breathing, diction, stance, presentation and form of subject matter' and 'practice speeches on set points of the Policy' (*Blackshirt*, 11 January 1935).

Blackshirt public meetings, whether in halls or on the street, were confrontational events and the BUF had initially forbidden women members from speaking in public. In early 1934, however, the first such meeting was inaugurated in Bromley, and the Women's Section organizing secretary, Mary Richardson, subsequently told *Fascist Week* that many requests for women speakers had had to be turned down because there were too few available (*Fascist Week*, 25 May 1934). By early 1936, however, the Chief Woman Propaganda Officer, Anne Brock Griggs, was reporting 'steady progress in the training of women speakers'. 'The old democratic parties', she claimed, 'could not produce speakers like these ... the majority young, courageous, yet level-headed, giving speeches which in their simplicity of language and sincerity could not fail to convince' (*Blackshirt*, 3 January 1936). Later in the year, working in the movement's East London heartland, Brock Griggs was in charge of women speakers who addressed crowds in Finsbury Square, Stepney, Bethnal Green, and elsewhere. Speakers like 'Mrs Brock Griggs, Miss Doreen Bell, Miss Olive Hawks, Mrs Carruthers and Miss Good', the *Blackshirt* reported, had brought about 'Record collections and paper sales ... accompanied by a gratifying flow of new members of both sexes' (*Blackshirt*, 12 September 1936).

Women speakers were also active outside London. In 1937, Olive Hawks was described as addressing a 'friendly and interested' meeting of farmers' wives at Whitstable and Doreen Bell as holding 'the interest of many of the women workers' at a factory gate meeting in Leeds (*Blackshirt*, 17 April 1937; 8 May 1937). Much interest, it was reported in 1935, had been caused by the 'experiment' of getting women speakers to open the ordinary Blackshirt meetings in Stevenson-square, Manchester, and the surrounding districts . . . and the large audiences which gather to listen are sufficient proof of the value of this introduction. Women's speakers' classes are being held regularly in Manchester, and before long it is hoped to send the ladies to Blackburn, Bolton, Preston and elsewhere. (*Blackshirt*, 21 June 1935)

Speakers' classes were not solely intended for BUF public events. In the late 1930s *Action* emphasized that District officials should not miss any opportunity to provide women speakers to address meetings of women's organizations while, earlier in the decade, a woman BUF speaker had debated 'Democracy versus Fascism' with a Women's International League representative at a meeting held at the Purley National Women Citizens' Association (*Action*, 18 June 1938; *Blackshirt*, 7 December 1934). In addition, much of the movement's effort to win over women was conducted less publicly. An early account of classes at Women's Headquarters describes many of those attending as being trained to make converts more informally and in early 1935 the *Blackshirt* urged those women members 'who spread Fascism privately among friends and acquaintances' to attend speakers' classes, where they would find their effectiveness 'enormously increased if they have practised their case clearly and impressively' (*Fascist Week*, 17–23 November 1933; *Blackshirt*, 1 February 1935). Two years later, in an article on the 'next stage in women's organisation', Brock Griggs noted the importance of afternoon meetings. These should be occasions in which 'invited guests should look forward to a pleasant afternoon every week when for an hour or so they will be among people who will become their friends'. A lecture on policy might be given or extracts read from British Union literature, and invitation cards, available from National Headquarters, could 'be given to any woman in her home who seems at all interested' (*Blackshirt*, 27 March 1937).

The subjects on which BUF women spoke ranged widely. In 1937, at one meeting in Limehouse, Olive Hawks spoke on the food supply, while at another, in Balham and Tooting, 'Mrs Selby spoke on the alien control of the British Press' (*Blackshirt*, 24 April 1937; 1 May 1937). But one recurrent topic was women under Fascism as, for instance, meetings held in Muswell Hill and in Harrow in 1935 evidence (*Blackshirt*, 29 March 1935; 31 May 1935). Nor was this topic restricted to special meetings. 'Again, and again', Olive Hawks wrote in 1934, 'at my outdoor meetings, the old question of women in the Fascist State is raised. Again and again it is said that in Fascist Britain women would be relegated to the home sphere, and all their newly-won rights taken from them'. This allegation, however, was denied, the BUF claiming that while Fascism would enable married women to stay at home, women would not be restricted from employment, would enjoy equal pay and would elect their own representatives to government (*Blackshirt*, 24 August 1934). I have argued elsewhere (Durham, 1990, pp. 5–8)[1] that BUF policy can be seen as ambiguous – its championing of equal pay, for instance, being argued sometimes as a matter of women's rights, sometimes as a way of making women more expensive and therefore less likely to be employed.

It remains the case, however, that part of the movement's appeal and part of its women speakers' armoury was the argument that British Fascism regarded both sexes as equal.

While some women were active as speakers, others took part in paper sales. In 1934 Leicester women members were described as selling literature in the streets daily, while one Brighton women's branch activist, Mrs Fife-Young, was reported to have sold 121 copies of *Fascist Week* in four days (*Blackshirt*, 15 June 1934; 1 June 1934). Another activist, Mrs Goodman, advised readers that she had found mornings and afternoons were the best times to sell the paper in Horsham: 'People sitting in cars parked by the roadside seem to be ready buyers. They have time to spare and evince a deep interest in the Movement when I approach them' (*Blackshirt*, 21 September 1934).

One form of activity which did not develop until 1936 was the Women's Propaganda March. Led by Olga Shore, the Woman Executive Officer of the organization, the first such march took place in May 1936, taking a route through Bethnal Green to Victoria Park. 'Women from shops, offices, and their own homes, donned their black shirts and marched through one of the most poverty-stricken districts of London', reported *Blackshirt*.

> And how they marched! It was a revelation to many who had doubted the ability of women to march properly. They marched in perfect step, with commendable order and discipline, the more praiseworthy because they had no band or drums to lead them. (*Blackshirt*, 23 May 1936)

This omission would eventually be remedied, an advertisement appearing in *Action* in July 1938 calling upon women members in the London area willing to learn to play the drums to contact the Woman District Leader for Westminster (Abbey) District, practice to be held every Tuesday at National Headquarters (*Action*, 2 July 1938). An appeal was also made to women members at the annual meeting of the Bethnal Green branch and in November the 4th London Area Women's Propaganda March gathered behind the Women's Corps of Drums, one week after its first public appearance at a march organized by the Central Hackney District of the movement (*Action*, 9 July 1938; 19 November 1938; 3 December 1938).

Of all the activities in which women were involved, the most important has not yet been mentioned. In 1934, one of the earliest accounts of women's work in the BUF had noted that 'without a sound knowledge of Fascism it is impossible for the women to undertake what is essentially *their* work in the Movement: that of canvassing'. This involved the typing of large numbers of index cards, dividing them into streets and blocks of streets and allocating these to particular members, whose duty it was 'to call at the houses regularly, distributing free literature, notifying the householders of meetings in the district, and answering the questions put to them' (*Blackshirt*, 1 June 1934). Mosley emphasized women's role in canvassing voters in a speech shortly after and Women's Headquarters immediately launched a canvassing drive in the Abbey Division, calling for at least fifty women to come forward. Any member, it declared, who could devote two evenings a week to attend training classes should apply. Fascist candidates, it urged, should be elected in every constituency in the next General Election, and while in the event none were put forward the following year, canvassing for future election to Parliament, for local electoral contests, and for general propaganda continued to remain important (*Blackshirt*, 6 July 1934). Just as Brock Griggs had emphasized this in the article in 1937 concerning the 'next stage in women's organisation', so, the previous year, another activist, Margot Cairns, had called for the battle to be taken to the homes. Many had attended meetings or seen marches but remained unsure; others had not come because they were busy or it had been wet or because 'they couldn't leave the baby'. Canvassing could reach them and women, she claimed, had more time and were more adaptable to the task. ' "Feminine intuition", "tact", call it what you will, there is no escaping the fact that women are the better canvassers' (*Blackshirt*, 27 March 1937; 11 April 1936).

The BUF did not participate in many local elections and while it selected a number of Prospective Parliamentary Candidates, the outbreak of war in 1939 ensured that the anticipated 1940 General Election never took place. Both nationally and locally, however, women were among British Fascism's candidates. Locally, Anne Brock Griggs was among its London County Council candidates in 1937 (emphasizing her hostility to 'Jewry', she stood in Lime-house, while another woman member, Mrs Warnett, stood in Shoreditch for the local council later in the year) (*Blackshirt*, 9 January 1937; *Action*, 27 February 1937; 18 June 1938). In Manchester, too, a woman member, Margaret Pye, ran for election to the council while, as for parliamentary candidates, eleven of the eighty selected were women, including Brock Griggs for Poplar (South), Olive Hawks for Camberwell (Peckham) and two more illustrious figures, Dorothy, Viscountess Downe, and Lady Pearson, for North Norfolk and Canterbury respectively (Rawnsley, 1981, pp. 185–6, 195; Cross, 1961, p. 179; *Blackshirt*, 20 March 1937; 12 June 1937; 19 December 1936).

Women's activity in the British Union of Fascists took many different forms and those discussed so far give us only part of the picture of women's role in the movement. As with other political parties of the period, much of women's activity in the BUF lay in the area of fundraising. In late 1937 Bethnal Green (NE) was holding jumble sales to raise money for women's branch funds while the Leicester Women's Section was building up the organization's election fund through whist drives (*Blackshirt*, 18 December 1937; 4 September 1937). Brighton, too, in September 1934, was organizing its women's section jumble sale (*Blackshirt*, 28 September 1934). If this was a highly traditional role, so too were the more menial tasks some performed. In 1938 Olive Hawks appealed for women supporters to join, suggesting that where some could sell publications or loan a room for afternoon meetings, others could be engaged in making goods for sale or 'the keeping clean and in order of District Headquarters Premises, which should not always fall upon active women members' (*Action*, 25 June 1938).

Other aspects of women's involvement, however, were very different. BUF meetings could end in violence and in 1934 it was reported that Head-quarters were offering women regular Monday classes in Ambulance and Bandaging, while the following year Manchester was described as having a growing women's section whose members were actively involved in First

Aid classes (*Blackshirt*, 5 January 1934; 8 March 1935). But there were other classes too. Where in Scotland in 1934 women members might be found at a Health Dancing Class, keeping themselves fit 'by eurythmic exercises', in Leeds in the same period they could attend physical training classes (*Blackshirt*, 27 July 1934; 15 February 1935). The Sheffield women's section offered a jujitsu class for women as did the South Shields branch, Manchester, and National Headquarters (*Blackshirt*, 12 October 1934; 16 November 1934; 8 March 1935; 5 January 1934). Perhaps more surprisingly still, at Women's Headquarters (and at the Carlisle branch headquarters) women attended classes in fencing (*Blackshirt*, 11 January 1935; 21 June 1935).

As the caption to a later photograph of 'Women Blackshirts ... fencing in Manchester' makes clear, fencing was favoured within the movement both because it assisted Fascist fitness and because it emulated Mosley's own chosen sport (*Blackshirt*, May 1938). Ju-jitsu, however, had a more specific purpose. The first public meeting addressed by a BUF woman speaker, in Bromley in early 1934, was attended by a Defence Force 'recruited from the local Women's Section' and under the command of the Women's Officer for the South East London area, Mrs Perry. The weekly meetings which followed were usually held in South London, Women's Defence Force stewards begin led by Miss M. Aitken, and an account of the time describes a training class at Women's Headquarters in which, in front of an audience of fifty young Blackshirt women, a young man demonstrated a ju-jitsu movement and asked two of the audience to ' "try to throw me out of the room!" There was a scuffle, and he was removed. "That's the stuff!" he declared, "now try it on each other" ' (*Fascist Week*, 25 May 1934; 17–23 November 1933). Aitken's force of women subsequently became known as the Propaganda Patrol or Special Propaganda Section, and in mid 1934 *Blackshirt* noted how 'similarly trained groups' were intended 'in every branch in the country' (*Blackshirt*, 1 June 1934). The following year a report appeared of the original group's first anniversary dinner at National Headquarters at which

> Miss Marjorie Aitken gave an interesting if somewhat unconventional account of what later became the S.P.S. It was interesting for those present to look back on the early days of the British Union of Fascists, and to remember how six women set forth to build up a Women's Defence Force.... From this nucleus has grown the most active branch of the Women's Section, the S.P.S. To-day they have both indoor and outdoor meetings all over London, held by their own speakers and stewarded by members. (*Blackshirt*, 3 May 1935)

That some women Blackshirts in the early days of the movement were trained in the use of physical force received national attention following in the violence at the 1934 Olympia meeting, when accounts appeared in the press not only of brutality by male stewards, but by women as well (Mullaly, 1946, pp. 45–6, 98). But if the early 1930s saw Fascist women using force, the late 1930s saw a very different form of activity. As war with Nazi Germany approached, British Union became increasingly involved in campaigning for peace. In October 1938 Brock Griggs had urged women to fight against what she claimed would be 'a politicians' war ... backed by Jewish money' (*Action*,

1 October 1938), and the following year, calling for meetings to be held during shopping hours, she remarked that

> Women speakers are particularly invaluable for such meetings . . . we reach mothers and housewives, not just political idlers and professional hecklers. It is these women who will suffer in this war, and it is also they who will swell the ranks of British Union in the fight for peace. (*Action*, 19 October 1939)

In January 1940 a Women's Peace Campaign was announced, to begin with a large meeting in February, followed by meetings 'all over London' in March and 'parallel campaigns' in the provinces. Money was appealed for from both genders to fund the Campaign's efforts to mobilize 'the wives and mothers of Britain' against the government's war preparations. London Districts set about planning their own Campaign Weeks involving meetings, paper sales, leafleting, and poster parades while from 'Hampshire, from Sussex, from Kent, from Lancashire, from Wales' came plans for similar activities (*Action*, 18 January 1940; 15 February 1940).

The London meeting, 'crowded to capacity with an audience composed almost entirely of women', was deemed a success (*Action*, 7 March 1940). The local activities which followed included Women's Peace Meetings at St Pancras, Tooting, and Ridley Road, Hackney, and a poster demonstration and leaflet distribution in the West End, and culminated, incongruously, in a women's meeting at the Quaker Friends' House, Euston, addressed by Olive Hawks, Mrs Booth (another of the movement's leading woman speakers), and Mosley himself (*Action*, 21 March 1940; 4 April 1940). At the meeting, Mrs Booth, who had also addressed a Women's Peace Campaign meeting in Manchester, illustrated her speech with 'moving instances from the life of the cotton towns' while another speaker, the former suffragette and woman's police organizer, Mary Allen, also appeared, her speech stressing 'the great part that women can play in the life of a resurgent nation' (*Action*, 11 April 1940; 18 April 1940).

This campaign, one of the last of the BUF's activities, is perhaps the strongest evidence of how diverse women's role in the movement was. But the organization of women Fascists was not without its tensions. In part, given British Fascism's male leadership and the views espoused by such figures as A.K. Chesterton, this is not surprising. But, as I have discussed elsewhere, BUF women themselves held divergent views on the role of women in Fascism, and where some, disconcertingly, argued that Fascism was compatible with feminism, others envisaged a future in which men would protect what was seen as 'the weaker sex' (Durham, 1990, p. 8; 1992, pp. 517–19). This combination of factors put severe strain on the BUF which its press hinted at but was loth to discuss, and it is particularly regrettable in this respect that no copies are known to have survived of the *Woman Fascist*, a penny fortnightly launched in March 1934. Edited by Elizabeth Winch, its first issue contained 'an article on Fascism and Religion, branch news and some practical hints for women fascists' and by June, according to a sympathetic press report, it was circulating to 'more than 700 subscribers' (*Blackshirt*, 23–29 March 1934; *Sunday Dispatch*, 17 June 1934). Marjorie Aitken, writing in the *Woman Fascist* in July 1934, urged upon women in the movement the importance of arguing that Fascism did not intend to 'force women back into the home' (*Blackshirt*, 20 July 1934). But if the journal was troubled by the scepticism of women towards Fascism, it was also concerned with relationships between the genders within the movement. In July the *Blackshirt* printed an extract from the *Woman Fascist* on 'co-operation between men and women for the triumph of the Fascist idea', remarking that it again brought 'to one's notice the co-operation that does exist between the men's and women's sections of the Blackshirt Movement' (*Blackshirt*, 6 July 1934). But the fact that this had to be so underlined is suggestive more of tension than of harmony and at the beginning of September, Sir Oswald Mosley's mother, Maud, Lady Mosley, now in charge of the Women's Section, announced that 'As has already been stated in the *Woman Fascist* . . . it has been decided in the interests of all concerned, to accept the *Blackshirt*'s kind offer of space devoted to Women's Section news. The *Women Fascist*, in its existing form', she declared, 'hardly represented the importance of the Women's section of the British Union of Fascists'. Every effort had to 'be made for men's and women's sections to combine as a team' and 'a "women's corner" in *our* paper' was to be preferred 'to any publication that might even suggest a separate existence' (*Blackshirt*, 7 September 1934).

Tensions flared again the following year when the decision to reorganize the movement to emphasize electoral work led to fears among some women that they would be forced to restrict themselves to canvassing, a belief that Mosley described as causing considerable difficulty for the movement (Lewis, 1985, pp. 67–8; *Blackshirt*, 22 March 1935). Women remained formally equal to men within the organization. But they were absent from its national leadership and even from District Leader level and the questions which the closing down of the *Woman Fascist* and the subsequent reorganization of the movement raised were effectively to disappear from its press in the years that followed. There was, however, a distinct echo when in early 1936 Brock Griggs was made editor of a women's page of *Action*. This, the *Blackshirt* announced, would be women's 'own little corner . . . mainly given over to the domestic side of life' but also dealing 'with general subjects from a woman's point of view'. It 'need hardly be said', the writer added, that 'Our women writers . . . will be treated on terms of exact equality with men, and will write on general subjects throughout the paper'. In this tortuous negotiation of a profoundly ambiguous view of women, the term 'it need hardly be said' barely veiled what precisely could not be said (*Blackshirt*, 7 February 1936).

The women who passed through the movement during the 1930s came from a variety of social backgrounds. Some – Maud, Lady Mosley; Lady Makgill; Lady Pearson; Dorothy, Viscountess Downe – came from the highest echelons in society. Other leading figures came from the middle class. Anne Brock Griggs was the wife of an architect; another Prospective Parliamentary Candidate, Sylvia Morris, was a freelance journalist and daughter of a doctor, while the Women's District Leader for Blackburn was 'the daughter of a well known Lancashire mill owning family' (*Blackshirt*, 9 January 1937; 28 November 1936; Rawnsley, 1980, p. 156). The rank and file stretched more widely. According to the reminiscences of the former Birmingham Women's District Leader, herself a teacher, 'Women members were . . . diverse, teachers, secretaries and office workers, nurses, shop assistants, waitresses, domestic workers, housewives etc.' (Brewer, 1984, p. 13). Calling for women canvassers in 1934, the *Blackshirt* had made 'a very special appeal' to the 'many women who have leisure in the daytime', which certainly suggests that some women Fascists were in more fortunate circumstances (*Blackshirt*, 6 July 1934). But writing of the Propaganda Patrol the same year, the paper noted how 'Many of these girls work during the day' and sold literature or attended meetings in the evening. As with the account cited earlier of 'Women from shops, offices, and their own homes' taking part in the first Women's Propaganda March, it is evident that the BUF was able to recruit from different strata of the female population (*Blackshirt*, 1 June 1934; 23 May 1936).

In order to win these women recruits, the Blackshirt press devoted considerable space both to its policy towards women and to the activities of its women activists. From early on, those who opposed British Fascism in the 1930s could not but notice that Mosley recruited from both genders. In early 1934 the feminist Winifred Holtby was moved to write an article on the subject after coming across a uniformed woman paper-seller in the street (Berry and Bishop, 1985, pp. 170–3) while the Communist Party's *Daily Worker* in the same period accused the editor of the *Fascist Week* of trying to 'soft-soap' a woman Fascist who had written in to his paper, expressing doubts about the movement's commitment to women's rights (*Daily Worker*, 7 April 1934). As the decade progressed, anti-Fascists continued to note the involvement of women in the BUF (see, e.g., Orwell and Angus, 1970, p. 230; *Jewish Chronicle*, 10 February 1939). But little attention was paid overall to women Fascists in the 1930s, and even less in the decades that followed. It is an omission that has long needed to be repaired.

Note

1   This essay criticized the neglect of both the BUF's pronouncements on women and the role of women within the organization. It was concerned primarily with the ideology of the BUF but also explored some of the reasons women activists gave for their involvement in the movement. In a subsequent essay, published in 1992, I have examined further how the movement envisaged the role of women in a fascist Britain and have discussed how BUF propagandists treated issues of gender.

References

BERRY, P. and BISHOP, A. (Eds) (1985) *Testament of a Generation: The Journalism of Vera Brittain and Winifred Holtby*, London, Virago.
BREWER, J.D. (1984) *Mosley's Men*, Aldershot, Gower.
CROSS, C. (1961) *The Fascists in Britain*, London, Barrie and Rockcliff.
DURHAM, M. (1990) 'Women and the British Union of Fascists, 1932–1940', in KUSHNER, T. and LUNN, K. (Eds) *The Politics of Marginality*, London, Frank Cass.
DURHAM, M. (1992) 'Gender and the British Union of Fascists', *Journal of Contemporary History*, vol. 27, no. 3, pp. 513–29.
LEWIS, D.S. (1985) *Illusions of Grandeur*, Manchester, Manchester University Press.

Martin Durham

MILLETT, K. (1971) *Sexual Politics*, London, Rupert Hart-Davis.
MULLALY, F. (1946) *Fascism inside England*, London, Claud Morris.
ORWELL, S. and ANGUS, I. (Eds) (1970) *The Collected Essays, Journalism and Letters of George Orwell*, vol. 1, Harmondsworth, Penguin.
RAWNSLEY, S. (1980) 'The Membership of the British Union of Fascists', in LUNN, K. and THURLOW, R.C. (Eds) *British Fascism*, London, Croom Helm.
RAWNSLEY, S.J. (1981) 'Fascism and Fascists in Britain in the 1930s', PhD, Bradford.

Chapter 10

British Feminists and Anti-Fascism in the 1930s

Johanna Alberti

Feminists in the 1930s belonged to a number of organizations, including different political parties, and there was no one central focus of thought and activity. Neither was there any one single Fascist view of women. The Fascist view as understood by the women in this study focused on the returning of women to the home, the controlling of the populace by propaganda which highlights the position of a leader, and the glorification of militarism. Many feminists of the period understood that behind those policies and practices lay a deep fear and even hatred of women – part of a general anti-feminist reaction which was itself part of a wider mood of despair. In the last weekly issue, on 2 October 1931, of *The Woman's Leader* (the successor to *The Common Cause*, the journal of the National Union of Women's Suffrage Societies), 'Crossbench' forecast an election, but despaired of the ability of Parliament to cope with the crisis and concluded with the view that the objects which the paper 'was founded to establish are not yet completely restored, and they will need a faithful guardian during the dark and difficult years that lie ahead'.

In the same vein Naomi Mitchison wrote in *Time and Tide* on 30 April 1932 that 'for civilization now the springs of life are failing. We are in terrible danger'. Ray Strachey's private letters also chart the personal and political process of a world 'collapsing piecemeal'. At the end of 1935 she felt that

> All that I can do for the moment is to go on tinkering with the present political machine and it generally seems to be worth while to do so. Now and then a sort of chasm opens up, and the whole thing turns futile.

Winifred Holtby referred in her contribution to 'Notes on the Way' in *Time and Tide* of 5 November 1932 to the 'period of reaction' as a journalistic platitude but acknowledged the contemporary loss of 'happy confidence in the march of civilization', which in turn had led people to cling 'with increasing fervour to accepted institutions'. She believed that writers had a particular responsibility in this mood of despair. In a review of *No Time Like the Present* in *Time and Tide* on 20 May 1933, she wrote in praise of Storm Jameson's condemnation of ' "La trahison des clercs de nos jours" ' – the treachery by which writers evade reality to satisfy the appetite for opiates . . . the political lethargy, the self-absorption, the lack of imagination which permits a civilization to drift towards disaster'. Virginia Woolf was later to use the image of the 'leaning tower' for the situation facing the writer in the 1930s. 'They had nothing settled to look at; nothing peaceful to remember; nothing certain to come'. (Woolf, 1940).

One reaction to this perception of the world on the brink of collapse was an extreme caution. In *Time and Tide* on 7 June 1933 Ellen Wilkinson complained about the predilection for the 'impartial' person, the fear of anybody with definite views. This attitude was part of what Winifred Holtby termed 'the Slump complex', characteristics of which were 'this narrowing of ambition, this closing in alike of ideas and opportunities. Somewhere a spring of hope and vitality has failed' (Holtby, 1934, p. 116). Within this context of fear it is possible to detect, as feminists at the time detected, both a reaction against women's achievements and a continuation of an older misogyny. The idea that feminism did achieve some of its goals during and after World War One has lost a good deal of credence today. However, some legislation in the immediate post-war period did make changes in women's political and social lives, changes which were perceived by contemporaries as significant. The perceived, or constructed, extent of these changes in turn led to an anti-feminist reaction of which Fascism was a part. Naomi Mitchison wrote of a 'counter-revolution' against the gaining by women of 'a certain amount of power' which 'has meant, politically, all kinds of filthiness and oppression'. (Mitchison, 1934, p. 104). And the historian Martin Pugh has described the 'resentment' of the 'parliamentary misogynists' which the first women MPs faced in the inter-war years, when, as Thelma Cazalet recalled, 'there was still something slightly freakish about a woman MP' (Pugh, 1992, pp. 190–1). The parliamentary correspondent of *The Woman's Leader* on 17 July 1931 compared procedure in the House of Commons to 'the traditions of the football field'.

It is possible to interpret the political equality gained by women as a mirage or as a smokescreen which disguised the reality of a continued misogyny rooted in a deep British distrust of women which is similar in origin to that strain in European Fascism. In a letter to *Time and Tide* on 7 November 1936, 'Honest Opinion' wrote:

> While Germany honestly and openly keeps women to the sphere which they adorn and refuses to allow them to meddle in affairs which they were never by nature intended to understand, England in practice quietly and unostentatiously pursues the same course – which is the only wise course – whilst paying theoretical lip-service to those ridiculous theories of equality which have helped to make modern democracy the laughable thing it is.

How then, in the context of a period of 'a general resignation by humanity of its burden of initiative' (Holtby, 1934, p. 116), did politically-minded feminists in Britain react to the phenomenon of triumphant Fascism in Europe?

Naomi Mitchison asserted in a review of Holtby's *Women and a Changing Civilisation* (*Left Review*, December 1934) that the effect of the rise of Fascism was 'of shoving a number of intelligent but (or should it be *and*?) politics-shy women into the field of political action'. Discussions at the Liberal Party Women's Conference in 1935 make it clear that some Liberal women certainly saw in the rise of Fascism a threat to the achievements of the Women's Movement: they also recognized the threat that Fascism posed to their political values as liberals. This sense that the Fascist attack on women was part of an attack on a wider system of values is echoed in a story Holtby tells of a friend who declares that there had been

> no 'rise of anti-feminism' in Europe.... There had been a rise of feminism; there is now a reaction against it. The pendulum is swing-ing backwards, not only against feminism, but against democracy, liberty, and reason, against international co-operation and political tolerance. (Holtby, 1934, p. 151)

The response of Labour women to the threat of Fascism shifted partly because of the expressed determination of the leadership not to be classified as feminists. Sometimes the emphasis was placed on the threat of Fascism to women, and at other times they apparently wished to avoid such an empha-sis. From the early 1930s, Labour women were clearly alert to the threat Fascism posed to women. The report of the fourth International Conference of Labour and Socialist Women in 1932 warned the delegates of the danger from Fascism to women 'in their world-wide struggle for freedom'. At the 1933 conference, however, the focus shifted to class: Barbara Ayrton Gould portrayed capitalists in Britain looking 'longingly at other countries which have reached their Fascist goals', and Ellen Wilkinson asserted that 'Fascism in Britain was coming through Lord Trenchard and his officer-class police force'. At the 1934 conference the lessons women drew from Germany and Italy were that the best defence against Fascism was the strengthening of trade union and Labour organizations.

Older and younger feminists sensed that their differing experiences led to different responses to the political context of the 1930s. In letters to her aunt, Elizabeth Haldane, Naomi Mitchison explored their different percep-tions of the place of women in politics, and in one she accused the older woman of continuing to separate the 'outside' from the 'inside' of life (Haldane papers, 1986). In *Poor Caroline*, published in 1931, Winifred Holtby portrayed the gap in experience between a woman of the 'unsheltered' older genera-tion, and a young woman who had not had to struggle in the same way for recognition. Molly Hamilton, who was a friend of Ray Strachey's and a Labour MP, believed that the woman of the younger generation was 'not interested in the status of her sex, as a sex. She takes it for granted that she be treated simply as a person, and resents everything different' (Strachey, 1936, p. 239). In April 1933, *The Woman's Leader* became *The Townswoman*, the organ of the newly formed Townswomen's Guilds. In the issue of October 1932 – one of the last issues of a feminist paper whose roots lay in the suffrage struggle – the book reviewer acknowledged sadly: 'Many of our younger readers know very little of the Suffrage Struggle, some of them have avowed that they are tired of hearing about it', but she added: 'To each generation its own "causes" '.

These generational differences may in subtle ways have undermined the resistance that feminists undoubtedly offered to Fascism. I now want to look at the nature of that resistance. For the pre-war generation one of the essential elements of their feminism was its emphasis on rationality. This construction of women was partly a reaction to the Victorian emphasis on women as intuitive and emotional and it did place them in direct opposition to the Fascist view of women. In her letters to her daughter, Ray Strachey emphasized the need for 'rational mental behaviour', and consistently constructs herself as emotionally stable, reasonable and unshaken by crises (Ray Strachey to Barbara Strachey, 6 November 1929; 26 January 1930; 19 February 1930; 26 January 1931). But in practical terms it was hard for women to hold it all together. Strachey was faced in the 1930s with the necessity to earn her own living and at the same time cope with some acute family problems. Although she remained very active as a feminist in the field of women's employment, her letters suggest that the focus of her considerable energies was narrowed from hope for the future to survival of the present: 'inching sideways through a tangle of temperaments'. She had few opportunities of 'sniffing the old suffrage agitation atmosphere again' (Ray Strachey to Mary Berenson, 11 February 1930; 28 July 1933). Winifred Holtby also believed that 'The enemies of reason are inevitably the opponents of "equal rights" ' (Holtby, 1934, p. 163). An admirer of her writing described it as 'sensitive, and humane, but also ironical, disciplined, courageous, and, above all, rational' (*Time and Tide*, 9 March 1935). Straddling two generations of feminism, Holtby also recognized that 'The problems which feminists of the nineteenth century thought to solve along the lines of rationalism, individualism and democracy, present new difficulties in an age of mysticism, community and authority' (Holtby, 1934, p. 7).

The Fascist construction of women was most energetically resisted by feminists in the first half of the decade. Feminists faced with the threat of Fascism were concerned to challenge any essentialist or constraining construction of difference, of 'masculinity' and 'femininity'. Winifred Horrabin thought that her contemporaries should be 'glad to be let off being men and women as such. After all, there are any number of "men" about, since Fascism is abroad in the land, and plenty of "women", see Hollywood' (*Time and Tide*, 25 February 1935). Ellen Wilkinson knew that the the 'platitudes about the difference between male and female nature' were invented to 'clothe the blunt fact that the girl on the machine can do the work faster than a man, that you do not intend to pay her for that superiority, and must therefore find subtle ways to prove that it is really inferiority' (*Time and Tide*, 28 September 1935).

Winifred Holtby's *Women in a Changing Civilisation* drew many feminist anti-Fascist threads together. She recognized the continuity of the struggle for freedom for women and the political threat of Fascism. She retained her faith that a new generation of women was aware of the threat to the fragile changes in women's status. She understood the connections between the Slump with its atttendant threat to 'women's right to earn' and the rise of Fascism. She saw how women's poverty and the fact that the education of a girl was still taken less seriously than that of a boy undermined the legal recognition of equality. Both in her own life and in her fictional writings she explored the difficult personal choices facing feminists. She knew that men feared women precisely because of the power of sexuality, that 'fierce magic, dangerous to men and yet essential to their comfort' (Holtby, 1934, pp. 6, 7, 94–5, 99, 67).

Margaret Rhondda, who worked closely with Holtby, was aware and was prepared to remind the readers of her paper of the depth of misogyny still prevalent, and of the potential for a Fascist treatment of women in Britain. Although she did not believe that the British Union of Fascists was a serious threat, she referred unspecifically in an article in *Time and Tide* on 21 April 1934 to 'other and more serious pretenders to Leader of the Planned State' who 'have this in common with Sir Oswald – that they share his views on the suitable position of women in that Utopia'. In this article she expressed her own doubts about women, suggesting that many would co-operate in accepting that position.

Lady Rhondda was most critical of women from her own class: she often evinced in her writing a powerful disdain of women who belonged to what she dubbed 'the parrot-educated stock begotten of the debutante system' (*Time and Tide*, 11 August 1934). But she did recognize that it was the social structures which were at fault. She described the system where men were proud that their wives did not work as one that made 'for almost unmitigated evil', acting on the women as a 'deadly poison' (*Time and Tide*, 14 November 1936). She also recognized that women's co-operation with Fascist ideas concerning the 'proper place' would stem from the fact that

> for all the talk about freedom, very few women have, in fact, achieved a position different in its fundamental essentials from that of their mothers. . . . A vote is all very well, but a vote doesn't make all the difference – ask the unemployed. (*Time and Tide*, 21 April 1934)

For middle-class feminists equality and independence involved a rejection of women's domestic role, a role emphasized and sanctified by Fascists. Rhondda recommended that men should share the domestic work, and Holtby delivered a punchy attack on sanctity of the home. She advocated that

> For the sake of human happiness, justice, intelligence and welfare, I should like to see all family homes and amateur housekeepers abolished for the space of one generation; because at present the price we pay for these luxuries is too high. We pay for them in humbug and irritation; we pay in narrowing down our interests and our loyalties; we pay in the deliberate stultifying of our human growth. Most of all, we pay in an orgy of vicious and enervating self-deception. (Holtby, 1934, p. 146)

Holtby was also critical of what she described as 'the cult of the cradle', particularly because of its failure to lead to practical recognition of what was needed to 'encourage willing and successful maternity' (Holtby, 1934, p. 168). At the beginning of the decade it had been possible for a feminist like Winifred Horrabin to construct an image of motherhood which directly challenged Fascism. She wrote to *Time and Tide* on 29 April 1933 in support of a letter from Ellen Wilkinson which had drawn attention to the treatment of women in Germany, and suggested that one way to draw attention to this and to protect the 'precariously held' liberties of women in Britain, would be to hold a meeting in the Albert Hall. In order 'to demonstrate to our embryo Fascists here that women intend to have a part in creating the new world', she wanted a protest that would 'express the strength, sense of humour and balance of an intelligent mother, as opposed to the stupid, cruel, dirty, little-boy type of mind which seems to be the moving spirit of Fascism'. This confident resistance, however, became muted as the decade wore on. In *Women in a Changing Civilisation* Holtby quoted a German woman who believed that Nazism had invested women

> with at least the same importance as men. The young generation obtain their first nourishment and teaching from their mothers. For this reason woman has again been recognised as the centre of family life and today it has again become a pleasure and an honour to be a woman. (Holtby, 1934, p. 157)

Honour in the home, pleasure in heterosexuality and the emphasis on family life were powerfully appealing to many women. To accept such an ideology was not to accept Fascism, but its acceptance could blunt the edge of the feminist response to Fascism.

Nevertheless there were feminists, and women who would not necessarily have defined themselves as feminists, who were alive to the danger Fascism posed to the autonomy of women. Women's independence was most strongly associated with the right to work for pay and feminists were aware that this was an issue where Fascism posed the most direct and obvious attack on women. Labour women were well aware of the connections between Fascism and attacks on women's work in the European context, although they did not overtly construct male opposition to their own defence of women's paid employment in terms of Fascism. There were undoubtedly people in Britain in the 1930s who would have been prepared to adopt Fascist policies concerning women's right to work. In September 1933, Sir Herbert Austin in a speech to a conference of directors professed admiration of Hitler's solution to the problem of unemployment: the removal of women from work in industry. Feminists were quick to see the implications of such views. Elizabeth Robins protested forcefully in *Time and Tide* on 30 September and linked this attitude with 'anti-semitism and other forms of hate and violence'. On November 14th 1933, a mass meeting was held 'to proclaim the right of married women to paid employment' (Holtby, 1934, p. 113). Labour women staunchly defended women's right to work: some included married women, although there was opposition to this. In the debate at the 1934 Labour Women's Conference, Miss Goodwin of the Association of Clerks and Secretaries made an impassioned plea that 'unemployment and other ills of the present system be rectified without taking from women their freedom to work as they desire'. She added that 'it was time that women realised the elements in the world today that were essentially Fascist. Women in Germany had been told "to get back to their kitchens" and this was making itself felt in England today'. In a letter to Edith How Martyn on 4 February 1938, Emmeline Pethick-Lawrence warned:

certain investigations which are being carried out with regard to the causes of the falling birth rate are only a preliminary to a determined effort to drive women out of industry and, that if we are not alert the Nazi policy towards women will be commenced by our own government.

In *Three Guineas* Virginia Woolf used 1936 issues of *The Daily Telegraph* to demonstrate the links between views expressed there about the employment of women and those of Hitler. She quoted a letter to the Editor:

I am certain I voice the opinion of thousands of young men when I say that if men were doing the work that thousands of young women are now doing the men would be able to keep those same women in decent homes. Homes are the real places of the women who are now compelling men to be idle. It is time the Government insisted upon employers giving work to more men, thus enabling them to marry the women they cannot now approach.

Virginia Woolf then commented:

There we have in embryo the creature, Dictator we call him when he is Italian or German, who believes that he has the right whether given by God, nature, sex or race is immaterial, to dictate to other human beings how they shall live; what they shall do. ... Place beside it another quotation: 'There are two worlds in the life of the nation, the world of the men and the world of women. Nature has done well to entrust the man with the care of his family and the nation. The woman's world is her family, her husband, her children and her home' [Hitler]. One is written in English, the other in German. But where is the difference? Are they not saying the same thing? Are they not both the voices of Dictators, whether they speak English or German, and are we not agreed that the dictator when we meet him abroad is a very dangerous as well as a very ugly animal? And here he is among us, raising his ugly head, spitting his poison, small still, curled up like a caterpillar on a leaf, but in the heart of England. ... And is it not the woman who has to breathe that poison and to fight that insect, secretly and without arms, in her office, fighting the Fascist or the Nazi as surely as those who fight him with arms in the limelight of publicity? And must not that fight wear down her strength and exhaust her spirit? Should we not help her to crush him in our own country before we ask her to help us to crush him abroad? (Woolf, 1938, pp. 60–2)

Another way of challenging anti-feminism and incipient Fascism was to remind women of both the achievements and the continuity of the struggle. Three hundred and fifty women at the Women's Freedom League Jubilee Dinner, as Vera Brittain noted in her diary on 14 May 1935, expressed a 'General feeling of satisfaction at progress made in 25 years'. The following year, *Our Freedom and its Results*, edited by Ray Strachey, reminded women of both the achievements of feminism and the struggle that was continuing. Four months before her death Winifred Holtby strongly asserted this continuity in an article in *Time and Tide* of 4 May 1935, contradicting those who referred to feminist demands as 'flogging a dead horse', and 'who say the world is no better off since "women have been let loose in it" '. Her attitude to such people was that 'they do not know what they are talking about'. She knew she had 'seen a revolution in social and moral values which has transformed the world I live in. It is a direct result of that challenge to opinion which we call the Women's Movement'. But she was also 'aware of the imperfections of the movement. I have seen what has happened in Germany, where the pendulum of reaction has swung back so violently that all that had been gained seems lost again'.

Winifred Holtby died in 1935; later in the 1930s it was harder for feminists to believe that their voices were being heard. On 4 September 1937 Helen Fletcher wrote in *Time and Tide* of her belief in 'the equality of the sexes' as something long gone; as for feminism, 'I didn't know any one remembered the word. It belongs to bicycles and Battersea Park'. Yet a month later, on 23 October, *Time and Tide* addressed itself directly to the question of what had happened to its own feminism when challenged by a letter from Monica Whately. An editorial response put forward the view that feminists should 'familiarise the general public with the idea of men and women working side by side regardless of sex and on equal terms for the general good', rather than 'perpetuate in the mind of the younger generation a picture of women as a class apart and inferior, always knocking outside the door, never doing, but always claiming the right to do'. Betty Archdale of the Six Point Group replied on 13 November with an appeal for feminists to do both; to belong to a feminist organization *and* to demonstrate their equality through work, 'showing in their lives that equality is a natural thing . . . the one without the other is useless'.

An appropriate image for what was happening in the 1930s might be that feminism and feminist discourse gradually sank just below the surface. The threat of Fascism, while it sharpened the response of some feminists, also shifted the focus for some. It is important in this context to recognize that the very real threat to the lives of European opponents of Fascism and their children meant that the specific threat of Fascism to *women* assumed secondary importance for many feminists. Those feminists who worked with refugees, or called attention to the victims of Nazism, were unequivocally opposing Fascism but could also be distracted from Fascism's specific attack on women.

The same was true of the international peace movement. The resistance to Fascism as nationalism and militarism was another vital political struggle of the 1930s which involved large numbers of women, including many feminists. In April 1935 Enid Lapthorn, who worked as Margery Corbett Ashby's secretary at the Disarmament Conference in Geneva, wrote in *The Liberal Woman's News* after a visit to Germany that she would like to write about the 'prevalent attitude to women' there but felt that war and peace were more important. Virginia Woolf would not have accepted this separation of ideas: her analysis in *Three Guineas* brought feminism and pacifism powerfully together. Similarly, the Women's Co-operative Guild reacted to the threat of Fascism by becoming 'increasingly pacifist' (Liddington, 1989, p. 159). Other feminists, however, took up a different ideological position on pacifism or else devoted themselves to practical activities, joining with men either in Popular Front movements or in work with refugees. Taking Ellen Wilkinson as an example – and her voice was heard consistently in protest against Fascism – as early as May 1933 she wrote an article in the *Gateshead Herald* drawing attention to the misogynist policies of the Nazi party. But her public attacks on Fascism were increasingly based on a class analysis of the political situation: she suggested in the House of Commons on 15 March 1939 that 'We have only to look at the attitude of many Hon. Members opposite to see that they regard the Fascist powers as being run by people who some-how stand for their own class'. In arguing thus Wilkinson was not necessarily implying that she had forgotten or abandoned a gender analysis of Fascism.

The anti-Fascist activities of feminists varied in scope and intensity. Virginia Woolf, for example, joined no organization, but *Three Guineas* demonstrates that she never lost sight of the threat which Fascism posed to women both within Britain and outside it (Oldfield, 1989). She wrote in her diary on 6 January 1935:

> And a teasing letter (the other night) from E[lizabe]th Bibesco.
> 'I am afraid that it had not occurred to me that in matters of ultimate importance even feminists cd. wish to segregate and label the sexes. It wd. seem to be a pity that sex alone should be able to bring them together' – to which I replied, What about Hitler? This is because, when she asked me to join the Cttee of the anti-Fascist Ex[hibitio]n, I asked why the woman question was ignored.

Monica Whately, as secretary to the Six Point Group, wrote to Woolf on 7 June 1935:

> Since returning from Germany at the end of last year, I have been in close touch with those working in Germany and for Germany in this country, and the reports received are more than disquieting. Every step is being taken by the German Government to close all avenues to women. She is being deprived of her right to higher edu-cation, of entering into the Trades and Professions – ruthlessly she is to be forced back into the home, the unpaid servant of her husband, but not the legal guardian of her children.
> What is perhaps more serious, is the fate of those women held as prisoners in criminal jails – not for any crime of which they have been proved guilty, but for the alleged misdemeanours of their men-folk.

Whately then described the fate of three such women – 'only three cases out of many'. The third and last was Mrs Friedel Worch who

> was arrested along with her daughter, a girl of less than seventeen, on July 22nd 1933. Mrs Worch took absolutely no part in politics. She with her young daughter was taken as a hostage because of her husband's politics. Early last October, Mrs Worch committed suicide.

Her daughter, who had just turned eighteen, was allowed to attend her mother's funeral, and was immediately afterwards sent back to camp. This girl has never taken any part in any political activity. She is a hostage and the treatment which caused her mother to take her own life, is still her daily lot.

In the hope that some of these women may be saved, will you take part in a Deputation of Prominent women to the German Ambassador? If we can make him feel that British women are deeply concerned at what is happening to the women of Germany, he may be willing to use his influence with the Government.

The degrading of women in Germany, lowers the status of women all over the world, and is bound to react against a better understanding between Nations.

Please let me hear from you as soon as possible for delay may mean, for some, we are too late. (letter to Virginia Woolf, 7 June, 1935)

The Six Point Group did not relax their efforts to alleviate the sufferings of victims of Fascism in Germany and Spain. In 1935 Monica Whately wrote a pamphlet published by the Six Point Group entitled *Women Behind Iron Bars*, drawing attention to the treatment of women opponents of Fascism. It was distributed by the Six Point Group which, together with the National Union of Women Teachers, was affiliated to the British Section of the Women's World Committee Against War and Fascism, although Monica Whately had her doubts about the commitment of the British Section to opposing the 'lowered status of women under Fascism' (minutes of the Executive Committee of the Six Point Group, 23 July 1936). This organization brought together communists, socialists, and feminists. In 1934 Monica Whately and the stalwart 69-year-old Selina Cooper were sent by the Committee to investigate German Nazism at first hand (Liddington, 1984, pp. 413–18).

Opposition to Fascism brought together women whose political views were not always in agreement. Rathbone, Wilkinson, Rhondda, and Lady Violet Bonham-Carter co-operated in trying to get a reprieve for Liselotte Hermann who was beheaded in June 1938 after she had tried to publicize abroad evidence of Hitler's rearmament of Germany for war (Oldfield, 1986, pp. 90–1). In April 1937, Ellen Wilkinson, Eleanor Rathbone, and the Duchess of Atholl travelled together to Spain and on their return all three worked devotedly to draw attention in the House of Commons to the evils of Fascism and to assist the escape of and bring relief to the refugees from those evils. Rathbone was especially cogent in her criticism of the failure of the British government to offer any help to 'men, women and children' who 'were enduring every kind of deliberately inflicted physical and mental torture'. She pointed out that 'nearly every receiving country has raised high walls with narrow, closely guarded doors' but the 'highest walls with the narrowest doors have been round Great Britain' (Stocks, 1949, pp. 258, 262). Not that those threatened with Fascism would always have chosen to be refugees: one of Rathbone's Czech feminist friends, Senator Plaminkova, refused 'the opportunity of safety. . . . She remained in Czechoslovakia, where in due course the Germans killed her' (Stocks, 1949, p. 261).

Many feminists found that the most practical action which they could take against Fascism was relief work with refugees, especially with female victims of German Nazism. Vera Brittain made 'innumerable telephone calls to the Home and Foreign Offices' on behalf of a Czech woman dentist after the crisis of March 1939. Her persistence was rewarded (Brittain, 1957, p. 197). Mary Sheepshanks helped organize relief for child victims of the Spanish Civil War, and 'gave asylum to a stream of refugees in her own home in Highgate' (Oldfield, 1984, p. 285).

Feminism was and is an international movement which stressed its links across national boundaries. In the inter-war period, Margery Corbett Ashby played a dominant role in the International Alliance of Women, as the International Women's Suffrage Association was renamed in 1926. She began her protests against Nazism in person:

> I had a dreadful lunch with the German delegation today. They asked for trouble by saying they hoped I was sending my nice boy to Germany in spite of 'the new Times', to which I replied that I certainly would not as long as there was no guarantee for any one's life or personal liberty ... I was shaking with nervousness but some one must tell them the effect on the outside world of their torturing and imprisoning Jews, Communists and Socialists. (letter to Brian Ashby, 25 March 1933, 483/C5)

Later that year she became active in the Liaison Committee of International Women's Organisations which sought to 'co-ordinate efforts that are being made to stem the set-back to the position of women under the new "Fascist" Governments' (484/C8). She began to work closely with Rosa Manus, a Dutch feminist active in the Women's International League for Peace and Freedom who was to be executed during the war in a German concentration camp. Corbett Ashby used her position as president of the IAW to protest against the the impact on women of the deteriorating international situation. In 1939 she pointed out in her presidential address that 'a world governed by force, brutality and fraud will find no place for women, save as breeders of men and forced labourers' (Harrison, 1987, p. 196).

The peace movement, a success in the sense of numbers involved, ended, of course, in failure. In October 1938, Naomi Mitchison wrote to Aldous Huxley in a state of despair: '*that* is what we with our antiwar propaganda have done, we've just made people afraid, so that everyone was enormously relieved to think that Nazi methods had won again' (8185). In contrast, Ray Strachey's letters at the time bear witness to her relief, to the feeling that nothing could be as bad as war. Maude Royden was one of many who joined the Peace Pledge Union after Munich, but by 1940 she had decided 'THAT NAZI-ISM IS WORSE THAN WAR' (Fletcher, 1989, pp. 273–4).

A final image of political impotence combined with personal commitment to individuals comes from the life of Catherine Marshall. A pre-war suffragist leader who turned to pacifism during World War One and worked for the WILPF between the wars, in September 1939, as war was breaking out, she was to be found worrying about the felling of trees on the island of Brandlehow in Derwentwater and concerned for the Czech refugees from Fascism whom she was sheltering.

I have argued here that feminist discourses challenging Fascism *were* constructed in the 1930s, but that they were robbed of their full impact by the circumstances in which they arose. Many women finally took their resistance to Fascism into practical activities which were anti-fascist, but not necessarily feminist. Fascism disrupted and muted the impact of such powerful discourses as were provided by Rhondda, Holtby, Wilkinson, Mitchison, and Woolf.

References

BRITTAIN, V. (1957) *Testament of Experience*, London, Gollancz.

FLETCHER, S. (1989) *Maude Royden: A Life*, Oxford, Basil Blackwell.

HARRISON, B. (1987) *Prudent Revolutionaries: Portraits of British Feminists between the Wars*, Oxford, Clarendon Press.

HOLTBY, W. (1934) *Women and a Changing Civilisation*, London, Bodley Head.

LIDDINGTON, J. (1984) *The Life and Times of a Respectable Rebel: Selina Cooper 1864–1946*, London, Virago.

LIDDINGTON, J. (1989) *The Long Road to Greenham Common: Feminism and Anti-Militarism in Britain since 1820*, London, Virago.

MITCHISON, N. (1934) *Home and a Changing Civilisation*, London, Bodley Head.

OLDFIELD, S. (1984) *Spinsters of This Parish*, London, Virago.

OLDFIELD, S. (1986) 'German Women in the Resistance to Hitler', in Reynolds, S. (Ed.) *Women, State and Revolution*, Brighton, Wheatsheaf.

OLDFIELD, S. (1989) *Women Against the Iron Fist: Alternatives to Militarism 1900–1989*, Oxford, Basil Blackwell.

PUGH, M. (1992) *Women and the Women's Movement in Britain 1914–1959*, Basingstoke and London, Macmillan.

STOCKS, M. (1949) *Eleanor Rathbone: A Biography*, London, Victor Gollancz.

STRACHEY, R. (Ed.) (1936) *Our Freedom and Its Results*, London, Hogarth.

WOOLF, V. (1938) *Three Guineas*, 1977 edition, Harmondsworth, Penguin.

WOOLF, V. (1940) 'Leaning Tower', a paper read to the Workers' Educational Association in Brighton in May 1940; published in *Collected Essays, Volume II*, London, Chatto and Windus, 1967.

WOOLF, V. *Diary* (Ed. A. Olivier Bell, 1982), vol. 1V, London, Hogarth.

Letters from MARGERY CORBETT ASHBY are in the Fawcett Library, City of London Polytechnic, London.

Letters from NAOMI MITCHISON are in the Haldane Collection in the National Library of Scotland, Edinburgh.

Letters from RAY STRACHEY are in the H.W. Smith MSS, Manuscripts Department, Lilly Library, Indiana University, Bloomington, Indiana.

Letters to VIRGINIA WOOLF are in her Reading Notebook/Scrapbook, *Three Guineas*, held in the Monks House Papers, University of Sussex Library.

Six Point Group Papers are in the Fawcett Library, London.

Chapter 11

Working with the '*Kindertransports*'

Veronica Gillespie

Kindertransport is one of those neat, German portmanteau words. It means a trainload of children. In the ten months between November 1938 and September 1939 it was a trainload of children being thrown out by Germany and accepted, with a few cautious reservations, by Great Britain. They came at a rate of one thousand children a month, nearly ten thousand in all. The Germans actually paid for the trains that removed these infant '*Untermenschen*' (subhumans) from the Fatherland.

Under the imperial splendour of German railway stations like Berlin, Dresden, Hamburg, Munich, etc. there would be a group of several hundred children well wrapped in warm travelling clothes, saying goodbye to heartbroken parents who had made the noble and terribly difficult decision to send them away because there was no future for them in Germany. They assured the children. 'We'll follow you in a few weeks', but most of those children never saw their parents again. Britain is inclined to congratulate itself on saving 10,000 children but at least three times that number whose names were already on lists for future transports were prevented from coming by the outbreak of war, and in all probability these went to the gas chambers. They were German, Austrian, and Czechoslovakian, plus a few Poles who had lived in Germany. About 80 per cent were Jewish and the rest Christians of partly Jewish ancestry, plus a few gypsies and others deemed unworthy members of the master race.

After the infamous *Kristallnacht* in November 1938, when Jewish synagogues, schools, orphanages, shops, and homes were reduced to rubble and broken glass, the International Finance Houses realized that they must raise a central fund for German-Jewish Aid. Prominent British bankers all contributed money and so did men and women important in Whitehall including the Foreign Minister at that time, Sir Samuel Hoare, a Quaker; Eleanor Rathbone MP; Ellen Wilkinson MP; Sir Wyndham Deedes; and Mr Norman Bentwich, a colonial civil servant who had worked in Palestine as Attorney General.

Women organizers in the British movement to help refugees included Jean Hoare, Sir Samuel's sister, and Elaine Blond (Mrs Neville Laski) who began as a fundraiser for the Children's Movement and then worked tirelessly in other departments for the rest of her life. Becky Sieff, her younger sister, and Lola Hahn-Warburg, sister of Kurt Hahn who founded

Jewish refugee children at UK customs

Gordonstoun, also worked for the young refugees, as did Mrs Dennis Cohen, who was in charge of refugee luggage.

In November 1938, soon after *Kristallnacht*, Mr Dennis Cohen, a publisher, was sent to Berlin to choose the children for the first *Kindertransport*: a horrible task as there were over 600 applications for 200 places. He allocated places to all the children from Berlin orphanages – he had to do that because the orphanages had all been destroyed on *Kristallnacht* – then places went to Polish children who were to be deported to Poland though the Poles refused to receive them. After that there were a few places which he gave to children from small towns where there was no Jewish community and no Quaker Emergency Committee since these children would be in the worst danger. After the first Berlin Transport in December 1938 it became too dangerous for an English Jew to travel to Germany and selection was then done by the Quakers.

Great Britain had agreed to accept an unlimited number of children travelling without parents, and also adults who, after a long, slow procedure with passports and permits, could come to take up a very limited range of work, such as unskilled hospital work, agriculture, or domestic service. There were also 'trainees' – young men and teenage boys from concentration camps, who, when they reached England, were housed in Kitchener Camp at Richborough near Dover, an old World War One embarkation camp, which they ran themselves on democratic lines. Every refugee coming to Britain had to be sponsored either by a British subject or by block sponsorship through one of the German-Jewish Aid funds. In total, nearly 10,000 unaccompanied children plus 60,000 others came, but it is not clear that the British government had fixed any limit on numbers.

In order to rescue the children quickly, Britain waived the need for passports for them. Instead they had identity cards stamped at the Home Office in London, then sent to Germany to be counter-stamped by the police, who always kept a few back for no known reason – just to show who was master, perhaps.

The identity cards, complete with photographs, were kept in the children's dossiers in Bloomsbury House, where I then worked. The photos were touching. Somehow each young face seemed to be saying 'I am a gentle, well-loved, lovable child'. And that was really true of most of them. The money for saving them was raised from many sources including the wealthy Jewish banking houses all over the world, the Save the Children Fund, Inter-Aid, and the Baldwin Fund. This was raised by Stanley Baldwin, a former Prime Minister appealing for funds 'not for some dreadful natural disaster, but for the disaster of man's inhumanity to his fellow men'. Half a million pounds was collected (equivalent in present-day values to ten million). Two hundred and twenty thousand of that came to the Children's Movement, to our Finance Department in Woburn House.

The money was needed for the children's fares. Although the German government paid for the actual *Kindertransport* trains, there were local fares to the main line, and also fares from Austria and Czechoslovakia. There was pocket-money of 10 Deutschmarks per child, the upkeep of hostels in England, the subsidizing of some of the placements with families (7s 6d per child weekly for food). And from March 1939 onwards the British Government asked for a £50 guarantee for each child. This seems mean, but it was because the original deal was that they would stay in Britain for two years only, and as the months of 1939 moved inexorably towards war, it was clear things would not turn out that way. After war had broken out, when the Refugee Children's Movement was running short of funds again, especially for the Farm Training schemes where boys were being trained for pioneering work in some new country after the war, the British government donated one million pounds and the USA two million pounds for training schemes.

It was of course useful to the British government to have a farm labour force of 15- to-18-year-olds working on food production once they could not pursue their usual policy of importing nearly all food from overseas. The farm-work plan was devised by Rothamstead Experimental Station, which was the main British centre for agricultural research, both for Britain and for her colonies.

Let us return now to the *Kindertransports* at the receiving end in Britain. The Quakers, the Jewish community, the various churches, and in some towns the trades unions and Workers' Clubs, set about establishing refugee committees in every large or medium-sized town in England, Wales, Scotland, and Northern Ireland, and in Dublin. These committees raised funds, obtained offers of Hospitality, prepared hostels, appointed Inspectors who would visit each refugee once a month, and were ready to receive children. There were sixty-five local committees in all, divided into twelve regions. It was at this stage that I became involved, though I was neither Jewish nor a Quaker. I had a German grandfather who had left Germany in 1865 to avoid conscription under Bismarck. I guess he was a pacifist. He died in 1932 and both my parents were also dead. I had an Oxford degree but did not want to teach and it was urgent that I should get a job. I was learning shorthand and typing at Pitman's College in Southampton Row, London, when I met Kate Gotthilf, a German-Jewish refugee. She spoke with a beautiful pure English accent which I later learned was the voice of Lady Violet Bonham-Carter to whose children Kate had been 'au pair'. She suggested that after our classes we would do voluntary work at a nearby office in Great Russell Street opposite the British Museum. Our speeds would improve more quickly, she said, taking down real letters instead of exercises. Besides, she was fascinated by a young Rabbi with his own aeroplane who flew to Vienna to rescue boys who would otherwise have perished in dreadful labour camps. I too was fascinated by this Scarlet Pimpernel scenario, though I thought it might have been a figment of Kate's imagination. As a matter of fact it was quite true and I think the young man must have been Wilfred Israel, the son of a wealthy Berlin family. He happened to have been born in England and with dual nationality could travel easily between England and Germany. He was shot down in 1941 over the Bay of Biscay after a mission to Spain and Portugal to help refugees stranded there. Quite apart from this, however, I was honoured to be asked to help in a project which proved to be one of the very best things I ever tried to do.

The rooms at 69 Great Russell Street were far too small and there were not enough typewriters. Early in December we moved to the shabby and empty Bloomsbury Hotel, and renamed it Bloomsbury House. Kate and I were offered paid jobs and £3 10s a week – a very good salary at that time. She was assigned to the Teenagers' Training Department at the top of the building while I was in the Aftercare Department of the children's section on the first floor.

Our accommodation was now roomy but Spartan. I came in the main entrance and punched a time-clock. Ahead was a huge reception-hall, by 9 a.m. already full of depressed adult refugees sitting in rows holding their numbered admission tickets, awaiting their turn to ask almost unanswerable questions or to draw their £1 15s a week subsistence allowance. You could live on that in London in those days – just!

I went up a handsome staircase to the first-floor offices of the Refugee Children's Movement. First door was the Record Room where the children's dossiers were filed under the care of a lady who had lectured in philosophy in Germany but who had come to England on a domestic permit (almost the only way a single Jewish woman from Germany could gain entry here in those days). She was working as parlourmaid in a luxury flat in Park Lane. Her employer was giving a luncheon when one of the guests, in general conversation, said 'Ah well, life is solitary, poor, nasty, brutish, and short, as Pascal remarked'. The parlourmaid plonked the man's soup in front of him saying 'Not Pascal: Hobbes'. As a result she lost her job and we gained an excellent custodian of records.

I was assigned as typist to Pam, a daughter of Lord Nathan, owner of Glaxo. (There is a statue of Pam, at the age of 1, over the gates of the Glaxo factory in New Zealand.) Together we used to answer 'aftercare' letters. One foster-mother was upset because her refugee child was better dressed than her own children. Another had asked for an orphan aged 3 and had been told there were no children available under 9 years old. (That first *Kindertransport* from the Berlin orphanages had given a false impression of the availability of a large number of babies and toddlers.)

The foster-parents making these complaints had been to Dovercourt Holiday Camp to collect their children. It was not a good distribution centre. The winter of 1938–1939 was one of the coldest ever recorded here and there was no way of heating the concrete chalets where the children slept, hugging hot-water bottles. They could use the dining-room, which had some heating, for meals, for English lessons, and for recreation, but afterwards they had to tramp through the snow to their isolated, freezing huts. Marks and Spencer sent a truckload of gumboots, which the children loved. Apparently these were not usual in Germany at that time. But this did not alter the fact that the sleeping accommodation was unsuitable and could not be properly supervised.

A woman called Anna Essinger had been appointed as supervisor of Dovercourt, and she now arrived there to advise. She was a very tall, heavily built refugee who formerly had taught at the highly prestigious New Herrlingen School in South Germany. Already in 1933, when people with foresight saw how things were developing, Anna Essinger had moved to England with all the pupils who were of Jewish or partly Jewish origin and set up a branch of the New Herrlingen in Bunce Court, a fine Kentish mansion with extensive grounds. She had the capacity to make children feel safe and sure of themselves. I always knew a Bunce Court child when they walked into my office. They were confident, well-spoken, and courteous. Most of the *Kindertransport* children were timid and mumbled in their effort not to offend anyone.

Anna Essinger went to Dovercourt, taking some of her staff and six members of her sixth form. She thought the place was dreadful and was appalled by the 'cattle-market' as she called the system by which representatives of local committees looked along the rows of children at meal times and picked out those they fancied. Plain adolescent girls or pimply boys were never picked, which did nothing for their morale. It was really cruel. Moreover Anna's assistants discovered that boys visited the girls' chalets at night. In the 1930s that was considered really wicked. Bright young things could sing Cole Porter's 'Let's do it, let's fall in love!' but teenagers destined for domestic work or agricultural training certainly could not! To add to the danger, these young people were in an extraordinarily tense and vulnerable emotional state. As Wordsworth put it:

> There is a comfort in the strength of love
> 'Twill make a thing endurable which else
> Would overthrow the brain or break the heart. (*Michael*)

It was no coincidence that a 15-year-old Polish girl once expressed exactly that thought to me in her own words.

Now the weather turned the rough grass in front of the chalets into a sea of mud and new *Kindertransports* were expected from Germany before places could be ready for them. St Felix School for Girls, Southwold, came to the rescue and gave the Refugee Children's Movement the use of their school for 200 boys for the Christmas vacation when the staff gave up their holidays to look after *Kindertransport* children, and a similar generous offer for the girls came from the Grammar School, Lowestoft. After Christmas, Anna Essinger managed to arrange that in future the children should go straight to London after disembarking at Harwich and should be handed over to foster-parents in the old staff gymnasium at Liverpool Street station. A rope was tied across the gymnasium with children on one side and foster-parents on the other. Two names were called out, one from each side, and the two met at the rope and went off together. How simple! But what a turmoil of thoughts and emotions in each of the two hearts!

What guidelines were used in deciding which child could go to which family? We did not bother about class but an attempt was made to match up religious background, Roman Catholic child to Roman Catholic family, Quaker or Lutheran child to Quaker or Protestant family, Orthodox child to Orthodox Jewish family and Liberal Jew to Liberal family. The problem was that there were nothing like enough Orthodox places, and even if a child's parents in Germany had declared that they would agree to their child living with a family of a different religion, Rabbi Schönfeld, the Movement's most influential Rabbi in Britain, would not countenance it. Many of the children, at this time of historic persecution, were serious about their religion. At St Felix School the boys asked if they might have a Yiddish Grace before meals and preface this with a minute's silence in which they could think about their parents. In the desperate effort to save as many children's lives as possible, many members of the committee perhaps paid too little attention to problems of religion.

It is said that Rabbi Schönfeld, almost single-handed, saved 750 children from Berlin and established them in Orthodox hostels in England, so he certainly did his share, but he was forever at odds with administrators on local committees who took the subject of religion more lightly than he did. 'What's better, a dead Jew or a lapsed Jew?' was their big question. But to Schönfeld a lapsed Jew *was* a dead Jew, or at least lost to the community. I do not know why offers of hospitality for *Kindertransport* children were slow in coming in from Orthodox homes. I think it was partly the very closeness of those homes. The parents knew that the stranger within their gates could change the balance of family relationships. They preferred to be generous with money (which they were) but cautious of giving parental loving care.

There were some excellent alternatives – the Schlesingers for instance. Dr Schlesinger was a senior physician at the Great Ormond Street Hospital for Children in London. He had four children of his own, including one who grew up to be the film director John Schlesinger. The doctor had recently inherited money which he invested in a house in Highgate. He furnished it to accommodate twelve children and paid a house-mother, a cook, and two assistants (all refugees). In fact small hostels for ten or twelve children were often the happiest places for refugee children because they were with others who could share their feelings and need not feel a duty to try to love a new father and mother. Then came the war. Dr Schlesinger was called up for the Army Medical Corps and all the children in the Highgate Hostel were evacuated to the country. It was very hard on the children to have to move for a second time, but at the end of the war the Schlesingers called them all together again and they had a big party to which they invited some German prisoners of war from a nearby camp.

Not all the placings were happy but most were successful. Out of nearly ten thousand children only fifty had to be moved to different homes or hostels. No doubt there were a few cases where the child was used by the host family as an unpaid servant, and even one or two cases of child molestation. It was very hard on the children when they had to be evacuated to the country because of the danger of air raids on big towns, and remaining faithful to their religion was difficult in the countryside where there were no synagogues. Some children asked to be baptized, to the horror of Rabbi Schönfeld. A correspondence course was started to try to keep the older Orthodox ones in touch. As for diet, the strict ones had to manage with vegetarian meals plus the occasional tin of kosher meat.

When I began work in Bloomsbury House most of the workers were young women rather like my boss, Pamela. But, as always with volunteers, there was a continuous shift. Every week a few left for other types of work. The Movement needed stability and suddenly Sir Charles Stead, who had been in the Indian civil service, was appointed Executive Director by the Home Office. He was not really suited to the job. Fortunately his Organizing Secretary, Mrs Dorothy Hardisty, was tireless, calm, and efficient. I myself saw very little of her when I was working in Bloomsbury House but I think one gets a good idea of her character from the following quotation from her Report on her work.

> The children had endured over a long period of time and in increasing severity such physical and mental suffering as had stolen their childhood from them. They were often old beyond their years, sometimes dreadfully experienced, always troubled and disturbed. It was not only that at short notice they were torn from the people they loved and trusted, and the places they knew: it was not only that they were suddenly bereft of all sense of security – these blows had been preceded by long periods of unhappiness and fear. Young children had seen the persecution of their relatives: men and boys being taken away from their homes: they had heard of the dread concentration camps. (Journal of Dorothy Hardisty – provenance unknown)

That comment comes from a woman of wide understanding. She was in her fifties when she joined the Refugee Children's Movement and carried it through very difficult years till 1948, when she retired at 65 and was awarded the MBE. She then worked for another twenty years running the Violet Melchett Infant Welfare Clinic near Sloane Square. I searched for a *Times* obituary of this remarkable woman – in vain.

Soon after Dorothy Hardisty joined the office, some of its 'here today, gone tomorrow' workers disappeared and we acquired Mrs Moore, mother of the Cambridge poet Nicholas Moore, and also Mrs Gaster, daughter-in-law of a former Chief Rabbi – a fine woman; I really loved working with her. Then I was moved from typing Aftercare letters and given the job of running the Re-emigration Department with a little supervision from Mrs Phipps, another newcomer. Our Chairwoman, Lady Reading, daughter-in-law of the former Viceroy of India, had a desk in Mrs Moore's room and came in nearly every day. Though at that time we did not know about the iniquitous 'ultimate solution' she felt sure that the refugee children faced a tough future and they should therefore not be pampered but encouraged to be tough and independent. Some of the other Bloomsbury House workers regarded her as too tough, but she was realistic and generous.

I was once getting a party of twenty boys who were emigrating to USA ready for their transatlantic journey. Some already had tickets, some had sponsors in England who guaranteed the single fare which was £20 by Cunard to New York or £18 by Furness Withy to Boston. But some had to get the passage-money from guarantors in USA who had promised to send it. One was very slow and the boy came to my office to concoct a telegram asking his uncle in New York to expedite the money. Back came the reply: 'Business bad. Ward will do all. Love and kisses, Abraham.' (I should mention that my surname at that time was 'Ward'). The boat-train was due to leave Euston the following morning. I rushed to see Mrs Phipps, my immediate superior. She suggested that Lady Reading might help, but Lady Reading had just left. I tore along the corridor, down the marble staircase and there she was! – standing on the front steps talking to someone from the financial office in Woburn House. I showed her the boy's identity card and the offending telegram as I panted out my story.

'What d'you say, Jim?' she asked the man beside her. 'Shall we go fifty-fifty?' They both signed the card in the 'payment' slot. On that kind of narrow chance a child's whole future could depend.

Quite a number of the children who were to emigrate had Nord Deutsche Lloyd tickets, which became useless as soon as war broke out. Others had Cunard, the America Line, or the French Compagnie Général Transatlantique with its beautiful ship, the *Normandie*, which was destroyed lying at anchor in New York. The Holland Amerika Line had two big ships, the *Volendam* and the *Veendam*. It was a dreadful tragedy when the *Volendam* was sunk by a German U-boat 200 miles from Liverpool where it was coming to collect many of our children, who were to join their parents on board and sail on to New York. There were few survivors. Many *Kindertransport* children lost parents in that disaster.

My routine now became one of calling at the Passport Office in Petty France, Whitehall, as soon as it opened in the morning, collecting any of the children's identity cards or passports that had been stamped with exit permits; then a quiet stroll to a travel agency in Jermyn Street to enquire about transatlantic passages.

The task was to find twenty children whose papers were in order for the USA and who had the necessary tickets or money, get passages for them through Mr Stacy of Hewitt's Travel Agency, and then bring them all to London, to our hostel in Lexham Gardens, South Kensington. I would go there in the evening for a simple '*Abendessen*' with them and make sure everything was in order. Then they would have a great sing-song of old Hebrew songs and psalms, plus perhaps some modern songs of their own – it was always moving, wonderful, and strangely reassuring. I would go home to bed, setting the alarm for 5 a.m., when I would cycle back to Lexham Gardens. How interesting London looked in the blackout, with just a bit of starlight or moonlight touching the barrage balloons that floated overhead. Nobody would be in the streets except perhaps an air-raid warden in his tin hat, and with the satchel for his gas-mask on his shoulder. This was the time of the phoney war – no air raids.

Taxis had been ordered at Lexham Gardens. Most London taxi-drivers were Jewish then, and they were always good to us. We would pack in and creep without headlights through the dark streets to Euston Station for the 7 a.m. boat-train to Liverpool. As I said goodbye to the children on the station platform, I was always gripped by fear of torpedo attacks at sea, but luckily all my parties got across safely – luckier than the British evacuee children on the *City of Benares* or the Italian internees on the *Arandora Star*. In addition to parties for the United States I sent smaller groups or single children to other places – the International Concession of Shanghai, Bolivia, Brazil, Peru. One boy went to Australia. I was not responsible for sending any to Israel – that was handled by the Youth Alijah.

Then came Dunkirk and after that the Battle of Britain. Young children from the London area who were not already evacuated to the country had to be sent to reception areas and there were areas near the coast which used to be considered safe but which now were in the front line for probable German invasion, so these children had to be moved too. Refugees over 16 years of age were interned on the grounds that they were 'disseminating alarm and despondency' – they were naturally very frightened, which was inevitable considering their experiences in Germany. Most of them were released from internment soon afterwards. Some were needed back at their jobs in food production, factory work, etc. Many who were 18 or over joined the army, at first in non-combatant roles with the boys in the Pioneer Corps and the girls as cooks, orderlies, or medical orderlies. Later, boys were allowed to join the fighting forces, and some went into the Paratroops. Thirty were killed in action and many were decorated.

The end of my *Kindertransport* job came in the autumn of that year, 1940. Groups of older teenage boys were sent from internment to Canada or Australia, but I was not involved in that project. For children still in the care of Bloomsbury House there were now no overseas passages. On 9 September 1940, at the height of the bombing of London, I was bombed out, so with no home, and no work for me at Bloomsbury House, I decided to join the army, and I spent the rest of the war operating searchlights around London and in Southern England. Before the war, like so many of my age, I had signed the Peace Pledge which stated that I 'declined to fight for King and Country' but if an enemy attacks your homeland I think it is right to defend it.

When the war in Europe ended in May 1945 all the personnel of the anti-aircraft batteries in Great Britain were required to attend a showing of a film about the liberation of concentration camps in Germany and Austria. There are no words to express the horror – the ground of trodden earth, the huge, dingy huts and the heap as high as the roof of the huts of human bodies, men, women, and children. Then an emaciated figure lurched out of one of the huts, a walking skeleton who moved jerkily towards the camera. The face should have registered despair or anger, but no, more frighteningly, there was a look of total vacancy. Oh God, I thought, don't let any of 'my' children see this! The agony for them was unimaginable. In early 1939, week after week, letters had come from parents in Germany, 'We expect to join you' and 'It won't be long now'. Then one came which said 'We are going on a journey and may not be able to write so often'. That had been the last, and now this film told us why.

Who would have thought, watching that film with horror fifty years ago, that 'ethnic cleansing' and child refugees would be agonizing life in Europe again, a challenge to British consciences once more?

Chapter 12

An Austrian Refugee in Wartime Manchester

Hanna Behrend

On 31 January 1939, aged 16, I was granted 'leave to land in Britain ... on condition that the holder registers at once with the police and does not enter into any employment other than as a resident in service in a private household'. I had come from Paris where I had spent an adventurous five months as a completely destitute refugee from Austria after my country had fallen to the Nazis in March 1938. The Anschluss scattered my whole family: first my widowed mother went to England as a domestic servant, then, early in 1940, she and my younger sister left for Ecuador. I only saw my mother once again, for ten days in 1955, and I have never seen my sister in fifty years.

Although I was under 18 – the official minimum age for an alien seeking domestic employment – I did find work as 'a mother's help' for 5s a week 'pocket-money'. But then, when war broke out in September 1939, I had to have special permission as 'an enemy alien' to be a probationer nurse in the County Mental Hospital at Prestwich. (A short-staffed mental hospital was the only place willing to take on an under-18-year-old enemy alien to complement their largely Scottish, Irish, and Black South African nursing staff.) However, the work permit was not issued promptly enough by the police, and so, just a fortnight after I had begun to learn the grim duties of a probationer mental nurse, I was sent packing again and not allowed to spend another night in the hospital. Where to go now for shelter and food? After a great many other charitable organizations had rejected my plea for support, the local Quakers took pity on me and placed me in a boarding-house that they ran for German and Austrian refugees in Manchester. I was there for two months, waiting for my work permit – just enough time to let me meet and fall in love with HK, the man who became my first husband.

My enthusiasm for this 28-year-old ex-student of chemistry from Berlin was grounded in my admiration for those who had offered resistance to the Nazis, sacrificed their careers or livelihoods, and even risked their lives when they (not being Jews) had had the option of conforming to the new regime's wishes and remaining unmolested. HK was a communist activist who had been tortured by the Nazis at the notorious police headquarters in the centre of Berlin. He had subsequently fled to Czechoslovakia; after the Munich Treaty he was lucky enough not to fall into the hands of the Nazi secret police entering Prague with the German occupation forces but to reach

England safely where he was supported by the Czech Trust Fund, a British government-sponsored fund for refugees from betrayed Czechoslovakia. He had left behind in Germany his parents and a girlfriend who had had a child by him. The girl committed suicide when the Gestapo arrested her, and friends in Dresden then took charge of the child. We tried to trace this child once the war was over but never found her. A silent, introverted, sombre-looking person, HK never joined the endless bickerings of the other refugee inmates of Mrs Smith's boarding-house. Neither did he get involved in the continual political arguments, except just once when he furiously attacked a Trotskyite for reviling the USSR. But he also – again in contrast with the others who were gradually becoming demoralized through lack of occupation, money, and hope – kept himself spruce, shaved daily, and never forgot his table manners.

What HK first thought of the '16-year-old', as I was then inaccurately called, I do not know. Our eventual intimacy was brought about by my initiative – I never missed going to places where I knew he was likely to be found and I suppose he just felt flattered and amused by the little fool's persistence. My very scrappy diary registers on 31 January 1940: 'Saturday afternoon off duty. It was lovely. I don't remember how it began but for the first time he really spoke to me as to a human being of some interest'.

By March 1940 HK and I were what in those days used to be called 'courting', i.e. we met regularly on my days or half-days off from the hospital and went to the pictures, to a dance, or for long walks eventually landing at a Lyons café for a cup of tea and a sandwich, or visiting friends together. Marriage plans were made. I was to take a commercial course on my free days and, as soon as I was competent enough, was to try and get an office job and then we would set up house together.

HK's experience of the betrayal of Czechoslovakia and the fact that Holland and Belgium had now fallen and the Nazis were advancing through France convinced him that the Nazis would invade England very soon. He was also certain that the Chamberlain government would intern the anti-Fascist Germans and Austrians so as to have them on the spot for the Fascist invaders as a kind of welcoming present. Although his forecast as regards the invasion of Britain, thankfully, did not come true, HK was proved right in respect of the internment of the anti-Fascist Germans and Austrians. The 'invitation' for HK came on 30 June 1940. On 10 July he was able to smuggle a letter out of Huyton-Liverpool internment camp in which he told me that there were plans to ship the internees overseas, a scheme which was only abandoned after one such boat carrying internees was torpedoed. The bulk of the internees were then taken to the Isle of Man, among them HK:

> This letter carries bad news. I am so excited that I can't think clearly. I have lost everything in the struggle against Fascism, parents, friends, health, etc., and England, priding herself on fighting for the liberty of all the nations, actually separates me with brute force from the only human being in the world that still matters to me.

It was to take nearly two years of lobbying senior academics, Quakers, and Eleanor Rathbone MP, before HK was acknowledged to be a proven, genuine anti-Fascist and finally released. The three hours I spent on the Isle of Man visiting him in that human zoo in November 1940 were the most harrowing experience I had in Britain. What shocked me most were the rows of ill-shaven, destitute-looking men behind the barbed wire fencing and HK's look of despair.

Meanwhile, at the County Mental Hospital, the period of the 'Blitz' meant that the staff had to be up and available outside their official shift hours any time that the siren went. On 30 August 1940 I wrote to HK: 'I have spent less than two hours in bed each night for the last week on account of the air raids so you can imagine my present mood'. When the sirens went we used to wheel bedridden mental patients and guide the others into various shelters and stay there with them until the All Clear. Naturally, whenever bombs dropped nearby and the building shook, there was literally bedlam – the poor patients, unable to grasp what was happening, were terrified by the noise and vibrations and screamed and prayed aloud. As there was a shortage of staff, I remember several such raids that I spent in a shelter, the only staff member, with some fifty mental patients. On 3 September there was another raid. I reported to HK:

I am just waiting for the bloody sirens; it is 10.30 p.m.; yesterday it was at 4.30 a.m. when I went to bed (for an hour and a half before the morning shift). This is why I am writing hurriedly.

1.30 a.m. on 8 September. I had to interrupt the letter because the siren went. I am dead beat. If I don't get a holiday soon I'll go nuts myself. I'll send you a parcel on my next day off.

On 9 September I was writing a letter when I was again interrupted by an air raid. At the end of October I left the hospital to work in a textile factory as a trainee machinist. But the Blitz went with me.

The bombing raid on 23 December 1940 was the worst Manchester suffered; I wrote in my diary:

Poor Manchester! I never realized how fond I had grown of this place until that Monday afternoon I won't forget so soon. Our works closed at 3.00 p.m. as nobody was doing any work, everybody having been up all night. I walked down St. Peter's Square when just as I was passing the corner building one of its remnant walls tottered and fell. We all ran and one poor AFS [Auxiliary Fire Service] man was buried under the debris, two men were saved. I will never forget their greenish pale faces when they, covered with dust and dirt, grabbed one another's hands ... so glad, so glad they were both alive.

Flames, fires, debris, ruins, Manchester has a new look these days ... one gets lifts in cars as there aren't either buses or trams. Lydia [a girl who had also lived in Claremont Road when I did] got killed. ... Aren't we fortunately built, we humans? How quickly we forget. How dreadful it seemed yesterday, today, we have not yet repaired the damage, but a mist has veiled our memory. Thank God for that!

For the paradox was, however shocking, that despite the war, the hard work the poor pay, and the poor food, and with my lover behind barbed wire, my young self still led a most active and even enjoyable life. Despite the very frequent air raids, I would spend nearly all my evenings and weekends at the Quakers' International Club, or with the Free German Youth or Young Austria, or at the so-called 'Tuesday class' – a social gathering run by a Presbyterian vicar and some of his church activists; and with my friend Heidi I had started studying Italian, French, and Russian. I was also an avid borrower of books from Manchester's wonderful Central Library.

Heidi, at that time a devout Roman Catholic, never missing Mass on Sunday mornings, would often join atheist me on hitch-hiking tours which took us to many places – for instance to Coventry the night of its Blitz, to Oxford and Cambridge, and to Birmingham. In August 1941 we spent our week's holiday walking in North Wales, staying at youth hostels which were cheap enough for us to afford them. I remember visiting Caernarvon Castle, walking through the fine woods of North Wales and enjoying it all tremendously.

During the working week, with a group of some ten young factory girls, English and refugee, we would go out regularly after our shift, to the opera, to see a Russian or a Polish ballet, to Prom concerts, or we would go swimming, or to a café. We always felt sorry for the married women with families who had to rush off home with never a minute's leisure and certainly never a chance of joining our weekend hiking tours. Occasionally, on Sunday mornings, Myriam, a Young Communist Leaguer, who also worked in my factory, asked me to join her canvassing for the *Daily Worker*. Both at the Young Austria Centre and at the International Club or at the Church social gatherings, debates would take place. I remember one on the subject of 'Should the law punish or prevent crime?' – I, of course, was the speaker for the latter view.

I stayed on at the factory until the end of August 1941, when, to my surprise, I was taken on as a clerical worker at Ryland's Export Company, thus rising in the world of labour from a manual to an office worker. I was thrilled, but scared. As I reported to H:

> My handwriting is so awful. A huge office boss looked at my application which I had tried to write as decently as I could and asked: 'Did *you* write that?' I: 'Yes, but I can do better.' (Ha, ha, ha, but also a lot worse!). My superiors believe I am proficient in Spanish and I therefore get all the Spanish letters to type but my Spanish is lousy. All the same I manage to cope. . . . I am now able to do my job properly. I can even work out in two ticks how much are $2598\frac{3}{4}$ yards at $7\frac{13}{16}$d and when I am asked to take a letter I no longer get the jitters. I know now that I can actually take down any letter without having to rely entirely on my memory, my good fortune and my wits. Can you imagine what it is like to have next to no knowledge of shorthand and no experience and somebody dictates a letter to you and all the time you wonder whether you'll be able to read back your scrawls.

Last week I started on a Russian and a Spanish course again. Did
you know that the Russians write a 'p' for an 'r'?

I was not always so jolly – at times the lack of sleep, the inadequate food,
the cold, the sirens, and the stress about H would all combine to threaten me
with a depressive breakdown. But nest-building, friendships and unlooked-
for kindnesses pulled me through. Elderly refugee women would cook real
Austrian meals for me, a Quaker lady treated me to a free country holiday
and I found a marvellous landlady who actually put pennies in my gas meter
in secret and refused to charge me for coal when she saw how little money
I had in my purse.

Finally, in March 1942, H was released and we immediately got married.
But marriage made no difference to my commitments. I did not even miss the
Young Austria Committee meeting the day after my wedding and the following week my diary listed four evening functions and two social gatherings!
We did, however, spend a wonderful week in May in a village in the Peak
District in the attic of a farm cottage. The farm labourer, who earned only
£1 a week, and his wife kept poultry and we enjoyed the rare luxury of farm
produce for our breakfasts and teas, went for long walks, and thus, belatedly
but none the less memorably, had a honeymoon after all.

As HK could not find a job in or around Manchester, he applied for a
vacant research chemist's job near Cardiff. And four years later, during the
bitterly cold famine winter of 1946–1947, I left Britain altogether, going with
HK to the Soviet Zone of Germany. I have been an East German ever since.

The three years I spent in Manchester were crucial for my development.
Not only had I met the man who was to play an important part in my emotional life; he was also instrumental in my political education. Though we
were separated for the bulk of that period he had initiated the process of my
becoming politically aware and contributed to my critical thinking. While he
was interned my experience as a working girl under wartime conditions and
the friends I found at work and in my leisure-time activities helped me to
grow up. I began to read avidly both fiction and history. My involvement in
the refugee organizations, particularly the German and Austrian youth groups,
but also in the Church group and the International club, sharpened my wits
and taught me a certain amount of political discernment. It was that critical
faculty which has frequently made me a square peg in a round hole and
for this I am truly indebted to my English upbringing. Unlike most of my
German contemporaries who had grown up under the Nazis, I was never
prepared to let other people think for me.

Dear Manchester! I have always loved you, black and sooty though you
then were, the way Lowry painted you, from the moment I set eyes on you.
Aye, duck, quite a lot of you has gone into the making of me.

Section III

Cultural History

Chapter 13

'A Fair Field and No Favour': Women Artists Working in Britain Between the Wars

Katy Deepwell

> In spite of such British personalities as the late Dame Ethel Smyth, the late Dame Ethel Walker, and the living painter Dame Laura Knight, women have not yet made, as women, any special impact on music and painting. (Vera Brittain)[1]

Vera Brittain's judgment, as she looked back on the inter-war years, was that only a few 'personalities' stood out amongst women painters (and musicians). These women, she implied, would be remembered as exceptions to their sex rather than as part of a distinctive contribution from women to the arts.

Vera Brittain's comments highlight what is a central problem for a critical feminist reappraisal of women artists working between the wars: the notoriety of a few women artists as 'exceptions' in contrast to the invisibility of the large numbers of women artists working in this period. Vera Brittain named two high achievers in the field of painting. Both were amongst the handful of women elected to the Royal Academy (RA) in the twentieth century and both were created Dames to honour their achievements as artists. Dame Laura Knight and Dame Ethel Walker were also, from 1932, the Honorary Presidents of the two largest women artists' groups in the country, the Society of Women Artists and the Women's International Art Club respectively. Each of these women artists' groups had memberships of 100–150 women artists and organized large annual open exhibitions half the size of the RA Summer Exhibitions.

As feminist art historians Griselda Pollock and Roszika Parker have argued, women have always been engaged in the production of art but in order to explain how women's marginalization is produced it is first necessary to examine how art history and criticism treats women's art production.[2] For it is just at the time when women were gaining greater economic and political freedom that women's cultural production has been progressively written out of the discipline of art history. The highest proportion of women artists in any study of this period, not surprisingly, is the most contemporary source and written by a woman. Mary Chamot's *Modern Painting in England* (1937)

Ethel Walker, *Vanessa* (1937)

includes 43 women amongst the 271 artists discussed (16 per cent). John Rothenstein's book *Modern English Painters: Sickert to Smith* (1952) includes only three women: Gwen John, Ethel Walker, and Frances Hodgkins are included amongst the twenty-two painters discussed – 14 per cent.[3] In Charles Harrison's *English Art and Modernism 1900–1939* (1981), amongst the 165 illustrations, the work of only three women artists is shown in ten illustrations (6 per cent). Women artists are mentioned only in passing and references to their existence can be found principally in footnotes. None of these texts discuss the work of the women artists' groups. Their invisibility in art history contributes to the understanding of women artists as isolated 'exceptions' in a male-dominated profession.

The marginal presence of women in art gallery collections or on public display today also contributes to this marginalization of women artists and the separation of women's art from 'Art' (male).[4] For example, 10 per cent of the Tate Gallery's collection of British Art is of work by women artists but less than 3 per cent is generally to be found on display. Of the 214 one-person shows held at the Tate between 1910 and 1986 only eight have been devoted to women (4 per cent).[5] The first of these one-person exhibitions was in 1935, and five of the eight have been in the 1980s.

Understanding how discrimination against women artists operates in art history and in representation in museums is an important part of understanding how many women's reputations have been buried and successful careers forgotten. There appears from these examples a form of cultural consensus about this period, that the appropriate level of women's representation as artists in both art history and public museums should be around 10 per cent. This level of representation is closely allied to art history as the history of modern movements, a history in which women artists are presented as minor followers and never innovators (see below). The figure of 10 per cent, however, remains a gross misrepresentation of the large numbers of professional women artists working in this period. The Censuses offer an indication of a broader picture albeit a necessarily highly qualified one.

Across the Censuses of 1911, 1921, 1931, and 1951, women figure as 25 to 33 per cent of all 'painters, sculptors, and engravers', of which the totals for both male and female were 12,498 to 16,548 in England and Wales, 1,089 to 1,599 in Scotland. Like their male peers, the women recorded are concentrated in major urban centres, e.g. London (in England and Wales), and Glasgow and Edinburgh (in Scotland). This professional category incorporates fine artists and commercial designers, craftsmen, their apprentices, and those working in small art and craft industries. It consistently excludes art students and art teachers in schools and colleges. Given the relationship between art teaching and fine art practice in the twentieth century, artists who teach could have been included as 'teachers' and not as 'painters, sculptors, and engravers'. The category does distinguish between those employed on their own account and those who were employers/employees. The numbers of 'painters, sculptors, and engravers' expanded throughout this period, due principally to the proliferation of industrial and commercial work, i.e., those recorded as employees or employed rather than those deemed 'working on their own account'. It is in the latter group that most exhibiting artists are likely to appear and in which the proportion of women was also higher than in the category as a whole, 31 to 39 per cent in the years 1921, 1931, and 1951 in England and Wales and Scotland (2,050–2,357 women in England and Wales, 110–206 in Scotland).

Another qualification is necessary in respect of the recording of the marital status of those in work. Single women appear to predominate, representing 66 to 83 per cent of those recorded amongst women 'painters, sculptors, and engravers' (2,969–3,434 in England and Wales, 190–270 in Scotland). By comparison, amongst all men in this category, single men represented only 35 to 48 per cent of those recorded. Feminist social historians have argued that the census classification significantly underestimates married women's participation in the workforce as casual, part-time, freelance, or temporary workers.[6] The under-recording of women's professional work is evident when one compares the Census figures to the numbers of known married women artists found in women artists' exhibiting groups, understood here as a major constituency amongst women artists, where roughly a third of their members in the inter-war years can be identified as married.

Contemporary feminists like Winifred Holtby insisted that women should be given 'a fair field and no favour'[7] in the workplace. What the Census figures interestingly suggest is not a picture of the continuous progress and expansion in the numbers of women artists but instead one of both expansion and contraction in the inter-war period. This is suggested (statistically) by examining the potential of all women in the population of working age who are recorded as working as 'painters, sculptors, and engravers'. This enables one to focus on women's opportunities compared to men's in the population given that there were 1,000,000 to 1,500,000 more women than men over school leaving age recorded in the population in England and Wales over this period. Of 100,000 men over school leaving age in 1921, 59 were 'painters, sculptors, and engravers' in 1921, 70 in 1931, 75 in 1951. Amongst 100,000 women 27 were 'painters, sculptors, and engravers' in 1921, 35 in 1931, and 25 in 1951. This form of statistics reveals a pattern amongst women as just under a third of 'painters, sculptors, and engravers' in 1921, rising to a third in 1931, and declining to a quarter by 1951.

The initial expansion of women artists can be seen as part of the expansion of the prominent exhibiting groups in this period. An increasing number of women were elected to artists' exhibiting groups up to the early 1930s. Women artists accounted for 25 to 33 per cent of the exhibitors at three of the largest mixed exhibiting groups: the RA, the New English Art Club, and the London Group, although the number of women exhibiting was always higher than the number who became members.

In this period women artists were elected to the RA for the first time since the RA's foundation in 1768 but they only represented 3 per cent of the Academicians and Associates by 1940 (rising to only 10 per cent today). Four women were elected as Associates during this period. All specialized in oil painting: the first was Annie Swynnerton in 1922. Two women, Laura Knight and Dod Procter, whose husbands were also Academicians, later became Academicans. In these circumstances, women artists who gained such a rare distinction saw themselves and were seen by the press as 'exceptions' to the general lot of women at home.[8] Feminists also sought to position the few women 'achievers' as role models for other women, noting their promotion in magazines like *The Woman's Leader.*[9] Inside the RA these women remained token figures. In 1936 when Dame Laura Knight became an Academician and was elected to the RA Council, a special statute had to be passed to allow her, because of her gender, to take her place. Women remained barred from the annual dinner for Academicians until 1967 when Gertrude Hermes protested and women were finally admitted.[10]

The four women elected to the RA did not, however, teach at the RA schools, where women represented just under half the number of students, before 1945. Prominent women artists, no matter how successful their exhibiting profile in the inter-war period, did not participate in the dominant model of art practice, that of the artist/art school teacher. In this sense, women artists were excluded from becoming 'gatekeepers', an American sociological term which refers to those in authority who have the power and cultural status to control who enters the profession.[11] While women artists generally emerged from the expanding secondary school system for girls initiated in the late nineteenth century, they also often returned to teach there as opportunities for employment in art schools remained so limited prior to 1945. Marriage bars introduced in the inter-war period reserved this lower-status teaching for single women.[12] Two good examples of women artists from this period who nevertheless became innovators in the field of art education are Evelyn Gibbs and Nan Youngman.[13]

The RA Summer Exhibition remained the largest forum for exhibiting artists in this period but analysis of the categories of work exhibited there reveals a pattern of discrimination in which the number of works produced by women cluster in the least valued categories: those from which the RA did not select candidates for election, like watercolours and miniatures (where women produced 95 per of cent of all exhibits). The RA principally elected Academicians who were oil painters and sculptors alongside a limited number of engravers and architects. In architecture, for example, women's representation was generally less than 5 per cent, and usually confined to displays of stained glass work. Women architects are not found amongst the individually named architects who displayed models of particular buildings. The first woman engraver elected Associate was Gertrude Hermes in 1963. Oil painting remained the largest category of work in the Summer Exhibition and a large proportion came from the Academicians and Associates. The proportion of women's work included did increase alongside the election of women to the RA and, in sculpture, as a result of the Lady Feodora Gleichen Memorial Award, introduced in 1926.

When women artists' work is discussed, discrimination against women artists is frequently presented in terms of 'obstacles', which individual women overcome by dint of strong character or feminist conviction: see, for example, Germaine Greer's *The Obstacle Race* (1979). As Linda Nochlin has argued, it is only through analyzing the institutions and education in which all artists, male and female, train and work that the operation of discrimination against women can be defined within the historically and culturally specific community in which they lived and worked.[14] The institutionalized nature of discrimination in training, exhibition selection, and different forms of cultural and social recognition in terms of honours, prizes, awards, and jobs has to be considered to create this broader picture. For these are the factors which

Edith Granger-Taylor, *Allegory* (1934). Reproduced from *Sunday Graphic and Sunday News*, 10 Feb. 1935, p. 19. Private collection. By permission of the British Library.

produce the differences between women artists' career patterns and critical reputations and those of their male peers.

Edith Granger-Taylor's painting *Allegory* (1934) and responses to it neatly highlight the problems facing women artists as seen by an artist whose career began in the inter-war period. In the composition, women are shown carrying satchels, signifying their talent, and pursuing umbrellas, which represent opportunities. The opportunities, however, are held exclusively by men. Described by Edith Granger-Taylor as a 'delicate feminist satire',[15] the artist's trenchant analysis in this work was clear. Women artists possess talent but lack opportunities. Their only route to success is literally to rise on the backs of men. *Allegory*'s reception highlights the way in which knowledge of discrimination against women artists, something that was part of women's experience in the inter-war years, fails to become part of public debate.

Although the press paid considerable attention to the painting when it was first shown in 1935 at the National Society of Painters, Sculptors, Engravers and Potters in London, they refused to debate its message. *Allegory* was described as a superlative 'modernist' picture 'because it is quite unintelligible'[16] and, more generously, as 'the latest problem picture'.[17] The press response amounted to a disavowal of discrimination, a refusal to acknowledge that any form of discrimination against women artists as a group existed. This approach was symptomatic of press attention to women artists and their work in the inter-war years and helps explain the presentation of a few women artists as exceptions. It forms the counterpart to the celebration of women as 'firsts' in the professions;[18] to the perception of their achievement as 'personalities' (Vera Brittain's words quoted in the epigraph to this chapter); and to women who are seen to succeed because they exceed commonly held assumptions about the weaknesses of their sex. It also reinforces the notion that success as an artist depends on individual qualities, on being a special kind of person who can transcend all or any adverse circumstances by force of personality alone.

Significantly, *Allegory* transforms the setting of Sandro Botticelli's *Mystic Nativity* (National Gallery, London) into that of an art school. Replacing the Virgin Mary and Jesus, a male art teacher at the centre of the composition patronizes a young female art student by laying his hand suggestively on her face. Enduring sexual harassment or being promoted through sexual liaisons with male artists, the work suggests, are part of the lot of the woman artist in an art school culture. Since the 1880s women students had formed 50 to 75 per cent of the art school population in Britain, although the number of women students varied considerably from art school to art school. At the Slade, for example, where Edith Granger-Taylor had studied, women students formed the majority of the students, outnumbering male students three to one. Life classes remained sex-segregated until 1945 and the teaching staff were all male with the exception of one assistant female teacher, Miss Alexander, employed from 1928. Gaining access to art education was not the principal problem that it had been for women who aspired to become artists in the 1850s, but the stereotype of the lady's accomplishment still persisted.

Women's development as artists was undercut by a training system in which they had to prove their sincerity as artists in the face of the assumption of peers, (male) teachers, and parents that they were simply 'matrix-to-marriage' girls. The autobiographies of women artists working in this period, for example those of Dame Laura Knight, Nina Hamnett, or Eileen Agar,[19] all make clear their individual determination to be taken seriously as artists in a culture which saw a career for them as professional exhibiting artists as unlikely and marriage as an end in itself. Dame Laura Knight's mother may have encouraged the ambitions of her daughter to paint but she enrolled her as a trainee art teacher with the idea that this would provide her, like her mother, with a secure income.[20] Teaching private classes or becoming involved in commercial design, mural painting, or industrial design alongside one's fine art practice were the other options artists, male and female, pursued to earn a living in the first half of the twentieth century. Women were often encouraged by career advice books and teachers to develop a craft specialism and it is significant that it was only the exhibitions of the women – only groups amongst the large fine art exhibiting groups in the inter-war years which provided opportunities for craft sections alongside the displays of fine art.

Edith Granger-Taylor's composition, while it places the art school experience at the centre of discrimination, also points to what happens outside this 'sacred' ground as equally important. Her own education included brief periods of study at the RA (*c.* 1910), the Slade, and St John's Wood School as a young woman, and in the early 1930s she joined Viadimir Polunin's classes on stage-painting at the Slade. She had by this time married, had a family, and established herself as an exhibiting artist with several one-person shows behind her. These experiences fed into both the language and subject of the painting.

In *Allegory* single women artists, however talented, remain stranded on the roof of the school. In contrast to the dominant ideology which saw marriage and a career as incompatible, Edith Granger-Taylor points to the dependence of women artists on men for opportunities. Family commitments for women have frequently been presented as an automatic handicap for women and not, as Barbara Hepworth describes it, as an enriching experience.[21] The negotiation of family commitments was/is rarely seen in a positive light, nor have domestic arrangements, including servants or nannies, been examined for the considerable time they gave many middle- and upper-class women artists to work and travel. The family as an automatic handicap for women also raises its head in the idea of the glass ceilings women set themselves or in the persistence of the 'internalised desire to fail' thesis.[22] Prime examples of such stereotyping from this period can be found in the biographies of Vanessa Bell and Dora Carrington, where an endless fascination with their sex-lives and domestic problems overtakes any interest in their professional activities.[23] Periods of self-doubt are explained by worries over love-lives or families.[24] Such explanations, particularly in biographies, 'concentrate on women themselves, rather than the structure, composition, recruitment patterns, market situation, and ideologies of occupations. . . . Domestic and family roles have been held responsible', and characteristics of women are seen as fixed and sex-specific.[25] Professional concerns over poor sales, patronizing reviews, or lack of professional recognition or promotion are rarely discussed as products of sex discrimination. Lack of economic or critical success, not managing to find an opportunity to exhibit on satisfactory terms, not having the physical space to work, an unsupportive partner or family, or the fact that little credence is given to their activities, all affect any artist's determination to continue, but with women artists explanations centre on family or sexual commitments.

A good example of the different opportunities given to male and female artists, drawing on the Slade School again, is the promotion of artists educated at the Slade into the ranks of the New English Art Club (NEAC). The links between the Slade and the NEAC are often discussed because of the prominence of male teachers and students who became members such as Professor Tonks, Frederick Brown, and Augustus John.[26] Comparative statistics highlight the limited opportunities for women compared to their male peer group as well as the general invisibility of women both as members and as exhibitors. Roughly half of the men who became members of the NEAC between 1900 and 1940 (46 of the 95 elected) were trained at the Slade, where they had formed only a quarter of the student population. Between 1900 and 1940 all but four of the sixteen women elected had been taught at the Slade (the four exceptions were Annie Swynnerton, Dod Proctor, Nadia Benois, and Winifred Nicholson). If the Slade provided the potential recruitment ground for new exhibitors and members of the NEAC, this trend was more marked for women than it was for men. However, this is not an indication of the triumph of the Slade as a qualifications lever for women artists because overall there is an inverse relation between their majority within the student population and their tiny minority within professional practice which actually reinforces the extremely low regard in which women students were frequently held by their male peers. For example, 20 per cent of the men elected between 1900 and 1940 had studied at the Slade in 1909–1910, yet the first woman from the same student group to become a member of the NEAC, Eleanor Best, was not elected until 1943.

Allegory also suggests that women artists' success is due to their relationships with men. A quarter of the women who were elected to the NEAC were married or related to male artists in the group. They include Evelyn Cheston, Grace E. Wheatley, Dod Procter, and Winifred Knights (Mrs Monnington); all but Mary Adshead (Mrs Stephen Bone), who was elected in the 1930's, were elected after their husbands. The NEAC's membership contained some notable single women known as 'the Cheyne Walkers' because of where they lived: Ethel Walker, E. Beatrice Bland, and Louise Pickard.

The 'advancement' of women within the NEAC, as with the RA, is also an extremely good example of the sociological trend in which women advance in recognized institutions as those institutions decline in critical importance. In 1888, when the NEAC was formally constituted, only two women were among its first fifty-seven members. By 1918 seven women were amongst the forty-five members (16 per cent) and in that year twenty-nine women artists were amongst the sixty-nine exhibitors (42 per cent). By 1940 their numbers had increased with fourteen women amongst the seventy-five members (19 per cent) but, although the exhibition was larger, the percentage of women exhibitors had declined with 60 women amongst the 153 exhibitors (39 per cent). Even where women shared the same education as their male peers, employed the same techniques, or worked and exhibited in the same arenas as their male colleagues, women were still seen (because of their gender) as being by nature incompatible with the profession's (male-defined, masculinist) vision of their own identity and sets of standards and values.

This is apparent in the very language which was used to evaluate work and placed women at a disadvantage in different interlocking critical frame-works. At the NEAC, artists were praised as 'true' or 'great'; sincerity and imagination were defined as the artist's greatest qualities. Women like Ethel Walker, by contrast, were 'distinguished', 'happy', or 'spontaneous'.[27] They were praised for their colour but not their facture, design, or composition. The subjects they undertook were trivialized and where they were discussed, it was to criticize their approach as unimaginative and derivative. At the RA, praise centred on a model of the academic naturalistic painter, and genius was defined by the possession of 'vision' or the revelation of 'truth' that a painter could inspire. Women were considered to have talent and skill. Women academicians like Laura Knight were seen as 'accomplished' academic painters but in art criticism the qualities of 'vision', and therefore genius, were found to be absent in their work.[28] At the London Group, where women like Ethel Sands and Vanessa Bell were members during this period, great emphasis was placed on 'originality' and 'being modern'. Women artists, at their best, had a 'personal' or 'individual' style but were considered to lack innovation.[29] At their worst, their work was derivative of all the qualities of established modern masters. 'Good' painting or sculpture by women was arrived at by accident and was not considered to be part of a process of sustained de-velopment or inquiry. Women in the London Group were often praised as colourists but condemned for their lack of design or plastic form, in the vocabulary of Roger Fry or Clive Bell. The same vocabulary and value system was adopted by the women artists of these groups, as Diana Gillespie's analysis of Vanessa Bell demonstrates.[30]

This spectrum of conceptions of the (male) artist against which the women artists within these groups are positioned became the means through which women artists' exclusion in art histories of this period, in museums, and as subjects for retrospectives has been justified as 'natural' and as a matter of minor interest.

Women had been increasingly successful in the inter-war period as members and exhibitors at the large exhibiting groups and many had had one-person exhibitions at regular intervals but few established sustained relationships with the new dealers in contemporary art who emerged in the inter-war years. The Slump and continuing effects of recession which were reputedly not felt in the art market until 1931[31] also had an impact on many artists deciding to take up other forms of employment. The early 1930s also saw a realignment of the art world with the impact of surrealism and abstrac-tion and the growth of social realist and politically committed artists as well as a switch from artists making their living through sales at exhibiting groups and occasional independently organized one-person shows to a dealer-led art market where artists exhibited with one dealer for sustained periods of time. Such changes in the marketing of contemporary art increased the level of discrimination against women, as can be seen in the level of participation of women in the Seven and Five Society. Women had formed a third of the members and exhibitors during what has been described as its 'lyrical phase' between 1926 and 1932; however, the numbers of women declined to 10 per cent as the group progressively became more abstract. At the Seven and Five between 1926 and 1932, another form of femininity was encoded in the critical language which was adopted to speak of the modernity of the group, but its vocabulary of 'fleeting delight'[32] was used to condemn women and praise men. 'Vitality' became the distinguishing term.[33] The rejection of the Seven and Five's 'lyricism' and calligraphy in favour of an 'art of pure form' as the group reorganized around the advocacy of abstract principles accompanied the departure of several women artists from the orthodox group that remained and established a new hegemony for form in the critical language of experiments in abstraction, over and above colour, space, or biomorphic abstraction. The change in the group's policy led to a change in gallery because dealers were unsure about possible sales. Those who left included Frances Hodgkins and Jessica Dismorr; those who stayed were Winifred Nicholson (exhibiting as Winifred Dacre), Barbara Hepworth, and the constructivist and critic Eileen Holding. Winifred Nicholson's interest in colour theory was marginalized by the experimentation with form in terms of carving and relief in the work of Henry Moore, Barbara Hepworth, and Ben Nicholson.[34]

In the British Surrealist group formed after the 1936 Surrealist Exhibition, women's participation hovered at that familiar figure of 10 per cent. As Whitney Chadwick has argued, women Surrealists negotiated an artistic identity in or around the (male) conception of the Surrealist muse and *femme-enfant*.[35] The rejection of what were effectively male projections about femininity encouraged women Surrealist artists to experiment with psychoanalytic procedures, automatism, or mysticism, but did not provide a basis for their re-incorporation into the 'official' movement and the 'official' account produced by men which continued to marginalize their contribution as followers rather than innovators. E.L.T. Mesens' attempt in 1940 to formalize the British group resulted in the exodus of several women Surrealists from the group, including Grace Pailthorpe, Ithell Colquhoun, and Eileen Agar (who was later reinstated).[36]

In direct contrast to these 'modernist' groups, the Artists' International Association (AIA), founded in 1933, developed a membership of artists and regular exhibitors of which 40 per cent were women. Though the AIA was run by a small group of left-wing and politically motivated artists with strong sympathies for social realism, the AIA from 1935 promoted itself as a professional body for all artists regardless of aesthetic allegiances. As it expanded the AIA significantly did not elect members as other groups did but defined their professional class of artist members through their art school training and participation in recognized forms of professional exhibition.[37] This procedure emphasized professional values without aesthetic allegiances combined with a political commitment against war and Fascism – an approach which attracted large numbers of women to join, in spite of the fact that its political stance eschewed feminism in favour of engagement with a male-defined and male-dominated left-wing agenda. Significantly, many AIA women members joined the Women's International Art Club in the 1940s and 1950s and were willing to identify themselves as feminists. Beryl Sinclair, for example, was Chair of both the AIA and the Women's International Art Club in the late 1940s.

The diversity of powerful cultural stereotypes used against women

Laura Knight, *Ruby Loftus Screwing a Breech-Ring*

artists within these artists' groups persistently and effectively homogenized women's work into a category separate from Art (a male cultural activity). This becomes particularly apparent in the reception of the annual open exhibitions of the Society of Women Artists and the Women's International Art Club in London. Their exhibitions reveal a fascinating and broad cross-section of women's art production in this period. Each contained in their membership and in their exhibitions diverse constituencies amongst women artists but though they identified their role as providing a platform or opportunity for women to exhibit, they did not see themselves as feminist organizations or as campaigning bodies for women's rights. An explicit feminist position, the creation of male critics, was used to harangue the women artists' groups in the press concerning the lack of any 'special' or 'distinctive' contribution from women to the visual arts.[38]

The impact of war also contributed greatly to the contraction in women's presence within the profession recorded in 1951. Nearly all artists had their work disrupted or redirected to the war effort between 1939 and 1945. Communities of artists which had existed before the war were disrupted and many groups and societies suspended for the duration. Women artists, however, did not gain more than 10 per cent of the war artists' commissions in World War Two and were mainly directed to subjects about women's work. In the post-war art market private galleries assumed increasing importance in the creation and maintenance of artists' reputations. Though the large artists' exhibiting groups continued, they became less significant in terms of generating an individual's critical reputation. The artists' groups that were founded after the war were considerably smaller. In 1945, the Arts Council was established and its new forms of state patronage which concentrated on distributing work through exhibitions did nothing to favour women.

In spite of an initial general advance of women in this period, women artists did not gain access to the important 'gatekeeping roles' of art school teachers, organizers, selectors, or jurors for exhibitions and commissions, museum curators, or mainstream art critics. The steady, but fragile, advance of women as exhibiting artists in this period was not buttressed through the appointment of women to these key roles and, therefore, was seemingly lost by 1951. Unless discrimination in the institutional context in which women artists live and work is examined alongside the gendered criteria of their 'success', the misogynist notion that weaknesses in women's character, destiny, or ability account for their lack of recognition will remain intact.

Notes

All statistics referred to in this chapter are drawn from my PhD 'Women Artists in Britain between the Two World Wars', Birkbeck College, London University, 1991.

1 Vera Brittain (1953), *Lady into Woman*, London, Andrew Dakers, p. 212.
2 R. Parker and G. Pollock (1981), *Old Mistresses: Women, Art and Ideology*, London, Routledge and Kegan Paul, p. 3.
3 See K. Deepwell (1987), 'The Memorial Exhibition of Ethel Walker, Gwen John and Frances Hodgkins', *Women Artists Slide Library (WASL) Journal*, 16 (April/May), pp. 5–6.

4   See Parker and Pollock (1981), pp. 8–9.
5   Figures from P. Barrie (1988), 'The Art Machine', *WASL Journal*, 20 (December/January), pp. 8–9.
6   J. Lewis (1984), *Women in England 1870–1950*, Brighton, Wheatsheaf, pp. 146–7.
7   W. Holtby (1934), *Women and a Changing Civilisation*, London, Bodley Head, p. 71.
8   See L. Knight (1965), *Magic of a Line*, London, William Kimber, p. 307.
9   *Woman's Leader*, 11 Nov. 1927; in Fawcett Library Clippings File: Artists.
10 G. Hermes, letter to H. Brooke at RA, 8 December 1966, artist's estate.
11 M. Bystryn (1978), 'Art Galleries as Gatekeepers: The Case of the Abstract Expressionists', *Social Research*, vol. 45, no. 2 (Summer), pp. 390–408.
12 A. Oram (1983), 'Serving Two Masters?', in London Feminist History Group, *The Sexual Dynamics of History: Men's Power, Women's Resistance,* pp. 134–48.
13 See K. Deepwell (1992), *Ten Decades: The Careers of Ten Women Artists born 1897–1906*, Norwich, Norwich Gallery.
14 L. Nochlin (1971), 'Why have there been No Great Women Artists?', in L. Nochlin (1989) *Women, Art and Power*, London, Thames and Hudson, p. 150.
15 Caption to photograph in *Sunday Graphic and Sunday News*, 10 Feb 1935, p. 19.
16 *Ibid.*
17 *Yorkshire Telegraph and Star*, 11 Feb. 1935; *Nottingham Journal and Express*, 8 Feb. 1935.
18 D. Beddoe (1989), *Back to Home and Duty*, London, Pandora, p. 75.
19 L. Knight (1936), *Oil Paint and Grease Paint*, London, Ivor Nicholson and Watson; L. Knight (1965), *The Magic of a Line*, London, William Kimber; N. Hamnett (1984), *Laughing Torso*, London, Virago; N. Hamnett (1954), *Is She a Lady?: A Problem in Autobiography*, London, Allan and Wingate; E. Agar (with A. Lambirth) (1988), *A Look at My Life*, London, Methuen.
20 J. Dunbar (1975), *Laura Knight*, London, Collins, p. 45.
21 B. Hepworth (1978), *A Pictorial Autobiography*, Bradford-upon-Avon, Moonraker, p. 20.
22 Matina Horner (1969), 'Fail, Bright Woman', *Psychology Today*, vol. 3, no. 6, quoted in R. Smith 'Women and Occupational Elites: The Case of Newspaper Journalism in England', in C. Fuchs Epstein and R. Laub Coser (1981), *Access to Power*, London, Allen and Unwin, p. 237.
23 F. Spalding (1984), *Vanessa Bell*, London, Papermac; G. Gerzina (1990), *Carrington: A Life of Dora Carrington, 1898–1932*, Oxford, Oxford University Press.
24 *Ibid.*
25 R. Smith 'Women and Occupational Elites', p. 237.
26 A. Thornton (1935), *Fifty Years of the NEAC*, London.
27 M. Chamot in *Apollo*, 13 May 1931, p. 307; Cora J. Gordon (1947) in *The Studio*, 133. Feb. 1947, p. 60.
28 See, e.g., C. Salaman (1932), *Laura Knight: Modern Masters of Etching 29*, London. The Studio; H. Osbourne (Ed.) (1979), *Oxford Companion to Art*, Oxford, Oxford University Press, p. 269.
29 On Ethel Sands see press reviews quoted in W. Baron (1977), *Ethel Sands and Her Circle*, London, Peter Owen, pp. 97–9; on Vanessa Bell see W. Sickert, 'Vanessa Bell', *Burlington Magazine*, 41, July 1922, p. 33.
30 D. Gillespie (1988), *The Sisters' Arts*, Syracuse, N.Y., Syracuse Press, p. 40.
31 O. Brown (1968), *Exhibition: The Memoirs of Oliver Brown*, London, Evelyn, Adams and Mackay.
32 H.S. Ede quoted in M. Chamot (1937), *Modern Painting in England*, London, Country Life, and New York, Scribner and Sons, p. 100.

33 See F. Hodgkins on vitality as a quality of genius, in E.H. McCormick, 'Frances Hodgkins: A Pictorial Biography', *Ascent*, Commemorative Issue, December 1969, p. 8.
34 Contrast Barbara Hepworth's and Winifred Nicholson's contributions to J.L. Martin, B. Nicholson, and N. Gabo (1937), *Circle: An International Survey of Constructive Art*, London, Faber and Faber.
35 W. Chadwick (1985), *Women Artists and the Surrealist Movement*, London, Thames and Hudson.
36 P. Ray (1971), *The Surrealist Movement in England*, Ithaca and London, Cornell University Press, p. 227.
37 AIA Application Form, *c.* 1942, AIA Archives, Tate Gallery, No. 7043.17.10.
38 See, e.g., 'The Society of Women Artists', *The Athenaeum*, 5 March 1920, p. 320; 'The Women's International Art Club', *Vogue*, vol. 55, no. 8 (Late April, 1920), pp. 70–1.

Chapter 14

British Women Surrealists – Deviants from Deviance?

Brigitte Libmann

No dream is worse than the reality in which we live.
No reality is as good as our dream.
(London Bulletin, June 1940)

In the Prolegomena to a *Third Manifesto of Surrealism*, 1942, André Breton stated that 'the relationship between men and women must be totally revised without a trace of hypocrisy'. The male Surrealists were dedicated to Mad Love; love was the only passion strong enough to bring down the frontiers between reality and possibility, truth and dream. Love enabled ordinary people to become poets and respectable citizens to turn into demons. Eroticism for them was not merely the cult of beauty, but a trigger for artistic creativity. Under the lash of the unconscious, sexual drives having been sublimated *à la* Freud, body and mind would reunite in an act of spontaneous creation. The basis of the philosophy of Surrealism was this connection of inspiration with spontaneity rather than with the will. The ability to let the unconscious express itself with no mediation other than the artist's hand was called Automatism. In the Surrealist process of creation there was no author, but a *scriptor*, a hand writing or painting under the direct influence of the unconscious. Like the medium coming into contact with the world beyond, the artist must relax the grip of reason and surrender to the flow of imagination. It was called 'disinterested thinking' because it was oblivious to all kinds of academic standards of conventional beauty. Beauty, for the Surrealists, came out of unexpected associations of ideas and images, revealing the existence of a 'law of chance' or *hasard objectif*. It brought to light a happy coincidence between the mechanical order of actuality and the realm of the highly improbable – the Absurd or dream world.

On account of this emphasis on spontaneity and the irrational, women and children were considered, theoretically, to be 'naturals' as Surrealist artists. But in actual fact it was very hard for women to become recognized artists, for they had also to outgrow the male Surrealists' dual stereotype of woman-as-child and woman-as-Muse. It has often been suggested that to become a woman Surrealist a woman artist had to become a male Surrealist's lover first (and indeed body-and-soul affinities are very often the quickest way into any avant-garde, secretive community). What is remarkable, however, is not the number of liaisons between male Surrealists and women artists – Eileen Agar with Paul Nash and Paul Eluard; Lee Miller with Man Ray, Cocteau, and Roland Penrose; Leonora Carrington with Max Ernst; Ithell Colquhoun with Tonio del Renzio – but that these women managed to assert their own artistic identity despite the inhibiting self-confidence of the more famous men. And the other British women Surrealists outside that charmed inner circle followed their own bent almost without male patronage or assistance of any kind. (On the contrary, as Emmy Bridgwater has testified, Roland Penrose excluded an obscure provincial like herself from any entrée into contact with the current work of his friend Picasso, among others.)

The International Surrealist Exhibition in London in July 1936 – a wildly controversial 'event' organized by Herbert Read, Roland Penrose, and David Gascoyne, with Dali in deep-sea diver's costume – was the catalyst that revealed the revolutionary possibilities of Surrealist art to sympathetic young women painters in Britain. Out of that Exhibition came the avant-garde periodicals *London Bulletin*, *Fulcrum*, and *Arson* to which many of them then contributed. This essay will look at some of the work of each woman individually and try to demonstrate in what sense it was 'surrealist' while also deviating from 'Surrealism'.

Eileen Agar

Eileen Agar, 1899–1991, had to rebel against her restrictive wealthy family in Argentina in order to become a working painter in Britain. Her clash against her mother's values she depicted in her surreal *objet* 'Hat for cooking bouillabaisse' – a send-up of her mother's extraordinary and extravagant turn-of-the-century headgear. This youthful rebellion sparked Agar's autonomy as she evolved her own synthesis of the representational principles she had learned at the Slade and the surrealist themes and techniques with which she experimented even before knowing that there was such a phenomenon as 'Surrealism'. Already before 1936 Agar had discovered on a beach an old anchor chain metamorphosed into another, new, 'natural' creation. This 'bird-snake' or 'seashore monster' may well have generated Eileen Agar's belief in the law of chance before ever she read about *'le hasard objectif'*. That was the first of her rich collection of *'objets trouvés'*. Because the re-creation of these objects comes from the passer-by's visual imagination, the titles are a vital clue to the artist's interpreting (sub)conscious. In her autobiography *A Look at My Life* (1988), Agar gives a sexual interpretation to the transfiguration of manufactured or discovered objects into intentional 'works of art'; she points out that 'the sea and land sometimes play together like man and wife, and achieve astonishing results'. In 1931, her life-long companion Joseph Bard and herself had founded a literary journal, *The Island*; she contributed an article on what she then called 'womb-magic', or 'the dominance of female creativity and imagination'.

> Apart from rampant and hysterical militarism, there is no male element left in Europe, for the intellectual and rational conception of life has given way to a more miraculous creative interpretation, and artistic life is under the sway of womb-magic.

The issue of gender and sexual representation would always be present in her later work, as indeed would her fascination with marine life.

In the chapter 'Am I a Surrealist?' in her autobiography, Agar said that she had always wondered if women were not indeed the real surrealist figures, because of the 'biomorphic' changes in their body when they are pregnant. Eileen Agar herself experimented with artistic creation as a metaphor for giving life, and for that reason did not want any children other than her work. The way she recalls her introduction to the crucial 1936 London Exhibition is interesting: 'One day I was an artist exploring a highly personal combination of form and content, and the next I was calmly informed I was a surrealist!' Agar makes it quite clear that she acknowledged Surrealism insofar as it gave her the opportunity to take part in a creatively stimulating social gathering. In fact, Picasso meant more for her than did Surrealism as a doctrine, perhaps because she outlived it. Despite her presence in the London Group and her contributions to the periodicals, she resisted the pressure to depict surreal disorder in her own creations. Her style is a worldly-wise mixture of decoration and abstract ideas.

Eileen Agar herself defined her work as 'half-abstract, half-surreal'. The way she explains the genesis of one of her most famous paintings, *Quadriga*, shown at the International Exhibition, reveals how the superimposition of imagination and reality worked on a conscious level for her. *Quadriga* is the transposition of a horse's head seen on a frieze on the Acropolis. The pattern of the head is repeated four times with variations. This composition is both classical, in its evocation of ancient archaeology, and new in the way the initial order is confused with the repetition of the subject and the random shapes produced by the unrestrained movement of the brush. Chaos and order together ensure the harmony of this film-like sequence. In the artist's own words 'it is like bridging a gap between the centuries'. This high level of conscious intention superimposed upon happenstance and intuition shows that she was never an Automatist – though she would not have said as much in the late 1930s.

As the world grew darker in those years so did Agar's work show signs of increased disquiet. Her second version of *Angel of Anarchy* (1940) – the 1936 version was stolen and never found again – is the plaster mould of a head, metamorphosed into an elegant monster, both attractive and frightening. It is a sort of ethnographic object, a wrapped-up female creature exhibiting her rich ornaments, whilst yet hiding her eyes and identity; a juxtaposition of heaven and hell, fashion and bad taste, peace and war – possibly in reference to the Spanish Civil War: Agar's commitment to the theory of Surrealism did not survive the Second World War. Perhaps she thought flippancy was not enough to make 'bullet-proof paintings' (cf. her own sardonically-titled *Bullet-Proof Painting*). Her elegant style could not protect her from the horrors of the London Blitz and her rationality kept her aware that art was helpless against Barbarism.

Leonora Carrington

Leonora Carrington, 1917– , almost a generation younger than her British women contemporaries, has had a life embodying the surreal elements of abrupt fissures, fabulous improbabilities, extreme cruelty and joy. She, like Agar, had had to rebel against a claustrophobic, constraining background of bourgeois wealth, aggravated in her case by a repressive Catholic boarding school education. From childhood she had determined to become a painter; her reading of Herbert Read and her visit to the International Surrealist Exhibition in 1936 was a revelation. Within months of having been presented at Court as a conventional society débutante, Leonora Carrington had eloped to Paris with the much older German Surrealist Max Ernst. There and in their rural hideaway at Saint Martin d'Ardèche, Leonora Carrington relished a new more vital world as she was introduced to all the Surrealist rituals and creative games. Reminiscences of her own fund of English nursery rhymes, Lear's *Book of Nonsense*, the fantastic world of Bosch, and the weird Celtic story-telling of her Irish Nanny all connected for her with the Surrealists' irrepressible, absurd, wild fantasies.

It was Leonora Carrington's private fascination with the esoteric and the occult, however, that led her to see the creation of art as a perfectly everyday process; her own witchcraft included writing with both hands and cooking extravagant, experimental dishes or painting her feet with mustard. Such doings were not just the product of her delight in practical jokes but also a kind of daily artistry. Like painting, writing, or sculpting, cooking should transfigure reality into magic. There is a rich but unstable ambiguity in Carrington's baroque images – the fairies, goddesses, witches, and above all the half-animal, half-human hybrids that haunt, always self-representatively, her pictorial world. One of her most striking self-portraits of 1937, *Femme et Oiseau*, depicts the artist as a horse. A humorous yet very frightening distortion of the self-image, it points to the desire to reverse the secular conception of woman as a powerless creature; the tiny black and white bird in the right corner of the canvas could easily be Max Ernst himself, a great painter of all that flies. The contrasting scale of the two creatures speaks for itself as to how the 20-year-old artist assumes *her* position.

Leonora Carrington worked at writing as well as painting her fantasies. One moving short story, *The Oval Lady*, written in French in 1938, is the tale of a 16-year-old girl, Lucretia, locked in a deadly conflict with her father who forbids her to play with her rocking horse; 'Let's pretend we're all horses', she says. This game of make-believe is the ultimate way to make dream-desires real. Indeed, Lucretia turns into a horse, thus gaining control over her father's threatening sexual power which is represented as an abstract, purely geometrical figure in opposition to Lucretia's humane oval face. In repression of this regressive game the father decides to burn the wooden toy. But the marvellous survives – 'the most frightful neighing sounded from above' – the horse – whether it is the girl or the toy we don't know – has a living soul, a soul for suffering.

Leonora Carrington's independence of vision induced loneliness. Indeed, her growing interest in the occult sciences progressively made her feel like a dissident within Surrealism. Dreams, according to her, could be read as transpositions of elements coming from a world beyond, whereas the Surrealists believed in the purely psychic nature of their oneiric visions, revealing the functioning of thought as an autonomous process.

Nevertheless, *Down Below*, an autobiographical account of her clinical

manic depression and her internment in a mental asylum in Spain in 1939–
1940 – before being rescued by her Nanny – this terrible narrative was praised
by the Surrealists as a great achievement. She was the first artist to come
back, successfully and sane, from a spiritual journey into madness. Though
she suffered a great deal, her account of her hallucinations as a temporarily
split personality, or Imago, was a break-*through* more than a breakdown.
Familiar with her overpowering unconscious, and armed with her now 'lucid
madness', she was able to go back and forth between reality and the sinking-
sand.

Such an extension of the creative field, through the total abandonment
of rationality, was one of the highest goals of Surrealism – as well as one of
the most risky. Thereafter Leonora Carrington continued to experiment freely,
moving between and crossing genders and species, ancient legends and personal myth. Her strange crayon and pencil drawing, *The Queen*, of 1945, for
example, is a vision of a lady-ship, possibly referring to the Celtic legend of
Tristan and Iseult. The inscription within the structure of the ship alludes to
the Queen of Ireland but beyond that are several other layers of meaning.
The female embryo/corpse inside the Queen is labelled '*Enfant interdit*' which
could mean both woman-as-repressed-child and woman-forbidden-to-have-a-
child (pregnancy being incompatible with the male Surrealists' conception of
woman-as-Muse, or indeed of woman-as-child).

Clearly Carrington's experience of 'Surrealism' was profound and real as
expressed in her highly idiosyncratic dream paintings. Even though she took
no part in the Surrealist Group's political gatherings and statements in the
late 1930s, she nevertheless managed to transmute many Surrealist preoccu-
pations into her own highly original haunted art.

Lee Miller

Lee Miller, 1907–1977, was yet another rebel, escaping, in her case, from the
family farm in Poughkeepsie first to Paris in 1925 and then to New York,
Paris, Cairo, and London. She was to make many breaks for freedom but the
most important and perhaps the most difficult was the break she made from
being a *Vogue* fashion model to becoming an experimental art-photographer.
There have not been many icons who have themselves become artists and it
is true that the images of Lee Miller are still better known than the powerful
images that she produced herself. Nevertheless those later images, some of
them the product of her chance discovery of 'solarization' when working with
Man Ray, have also proved immortal.

Self-portraits are crucial to an understanding of the artist's perspective
on her own work and personality. Self-portraiture was especially important
for someone whose body and face had been objectified and alienated by the
spectator's glance. For someone like Lee Miller, therefore, self-portraits hold
a double meaning, both personal vision and projection of what was expected
by the other's gaze. Her *Self-Portrait, 1933* is at first sight a quite formal picture,
presenting the artist's classical profile, dressed in velvet against a black background. It could be just a fashion photograph – until one remembers that the
eye of the one portrayed was also the eye behind the camera. One then

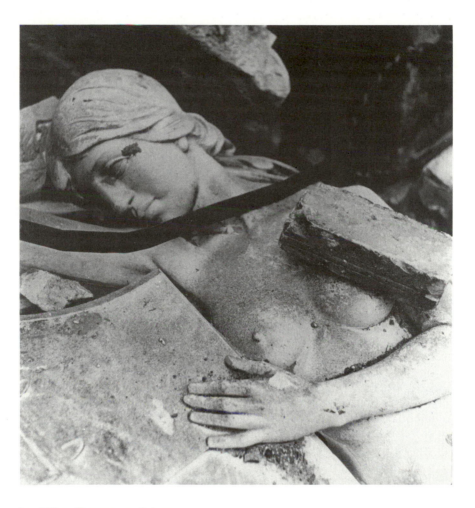

Lee Miller, *Revenge on Culture*

realizes that the model is looking away from her (self) portrayer, possibly into a blazing fire, since her chair and face are lit up by a crude white light that contrasts with the dark Renaissance dress and black velvet gloves. The real subject is actually decentred, the eye is attracted to what it cannot see – the invisible source of light. What most disturbs the apparently formal portrait is the expression of mingled melancholy, nostalgia, and disillusion on Lee Miller's face, for a model is usually presented as an emotionless blank.

Between 1932 and 1937, Lee Miller experimented as a photographer in Paris, New York, and Cairo. After linking up with Roland Penrose in 1937 she became the unofficial photographer of the Surrealist Group, chronicling their abandonment to holiday and many a *déjeuner sur l'herbe* enjoyed by herself and Roland Penrose with Eileen Agar and the Eluards, or Leonora Carrington and Max Ernst, Picasso and Dora Maar. These pictures, many of them unpublished, are held in the Lee Miller Archives, Chiddingly, Sussex.

World War Two of course destroyed that idyll, but whereas the war ravaged Agar's creativity and Carrington's sanity, it inspired Lee Miller to produce her most significant and memorable Surrealist images, summed up in the title of her book of photographs, *Grim Glory* (1942). Her picture *Fallen Angel 1940,* also known as *Revenge on Culture,* is a tender, symbolic evocation of all the anonymous victims of man's destruction. The 'angel' is a female statue lying on her side in the rubble among tombstones, the victim of an air raid. A deep cut, made even darker by the developing process and looking like a black strip of metal, severs head from body, suggesting that here is not an accidental death, nor a natural one, but a murder. The blind look of the dead statue's gaze is in painful contrast to the movement of her hand, tenderly protecting a large book, as if, after her fall, she were still trying to save culture from/for humanity. In her martyrdom she has lost the rigidity, the untouchability of stone; in her incriminating surrender to violence she has become a political symbol of peace – instead of the merely erotic symbol of lust that the statue in Bunuel's *Age d'Or* (1932) had been.

Lee Miller's direct participation as a photographer among the Surrealists was short and ended with the war's end, when she reported as realist, not Surrealist, what she found on entering Dachau. It would seem that finally Lee Miller could not share the Surrealists' sense of derision and mockery. Rather, she embodied the strong, rebellious woman of the 1930s, so involved in her own search both in life and in art that she did not have time for a group that kept asking questions but never came up with answers.

Eileen Agar, Leonora Carrington and Lee Miller were, however intermittently, part of a cosmopolitan inner circle. The other British women painters associated with Surrealism worked exclusively in Britain and had little or no such intimacy with the 'stars' of the European avant-garde. Each of them found her own way into, and in many cases out of, Surrealism as a movement.

Grace Pailthorpe

Dr Grace Pailthorpe, 1883–1971, was the most unexpected person to find among the Surrealists yet her journey towards them had its own logic. A doctor who had served as a woman surgeon in the slaughter of World War One, she had become, as had Breton and his friends, a passionate pacifist. She turned to the new theories of psychoanalysis as a mode of healing, eventually becoming influenced by the work of Melanie Klein and pioneering a call for the psychiatric treatment rather than the criminal punishment of juvenile delinquents – especially girls. In 1931 she founded the Institute for the Scientific Treatment of Delinquency, today the Portman Clinic, and in 1932 she published *Studies in the Psychology of Delinquents* and *What We Put in Prison*. Pailthorpe was interested both in Automatism and in the art of the insane, being one of the first practitioners of art therapy. In 1935 a young artist, Reuben Mednikoff, became her assistant; she started analyzing his paintings and drawings and he hers. They presented their work to the International Surrealist Exhibition committee in 1936. Three of their works were selected, a watercolour and two ink drawings. Another watercolour of Pailthorpe's was exhibited in 1939 at the Zwemmer Gallery: *The Torment of Tantalus* is a spontaneous drawing, and seems to be made in one go, as in a non-stop line, except for a more architectural feature, reminding us of the presence of rationality overshadowed by the apparently haphazard scribble. The result is a mixture of children's art and cartoons. There is of course a humorous clash between the drawing and its classically referential title; Tantalus, a Homeric hero, was punished after he told the other mortals about the secret of the Gods. The punishment itself differs in the many versions, but epitomizes here the ominous, tortuous guilt complex in man. This drawing is typical of what the critics present at the International Exhibition in 1936 called depreciatingly 'serious Nonsense': there is nothing to it, apart from an endless interrogation of what the punishment was, and the fantasy about it by both artist and spectator. Pailthorpe herself declared defensively: 'In the process of becoming free, surrealist painting, drawing or sculpture will necessarily be infantile in content. This does not preclude its right to be called Art'.

But the irony about Pailthorpe's own work lay in her *un*intentional humour and the solemnity of her theoretical views concerning the naive. In January 1939 she published a provocative long article for the *London Bulletin*, No. 7, 'The Scientific Aspect of Surrealism', in which she tried to prove that the final goal of surrealism and psychoanalysis is the same – the liberation of man. The fantasy story that the unconscious is unfolding is intelligible. In other words, Pailthorpe tried to appropriate the Surrealists' intuitive link with the subconscious for an objective science of the psyche. But her own interpretation of Mednikoff's daydream-cum-automatic-drawing entitled *Come Back Soon* is both deeply subjective and wildly, unconsciously, funny. From this image of a horse-headed creature with a rooster's cock's-comb running on two male legs behind which lies a box holding (possibly) a dead bird, Dr Pailthorpe is able to deduce:

> The underlying unconscious fantasy is that the artist has killed his mother and is now enjoying himself with playing with the mess the kill has provided for him. To do this he has first to decorate himself with a cock's-comb and the beard. . . . In his childhood he has witnessed 'kosher-killing' of poultry. Priests with beards become to his

> child-mind *the people who may kill*; therefore, in his fantasy, he first
> makes himself into a priest with a beard. . . . The dead bird repre-
> sents the mother [to whom he says 'Come back soon'].

Such a commentary is hardly a scientific analysis but rather is a classic absurd
Surrealist text. Grace Pailthorpe and Reuben Mednikoff were accused of
trying to subvert the very doctrine of Surrealism, threatening its autonomy in
their utilitarian approach to Automatism for psychological investigation.
Allowing Automatism to yield up medical and socio/psychological findings
was to allow society's institutions to get hold of it – libertarianism would risk
being hijacked. Paradoxically the orthodox Surrealist insistence on anarchic
unorthodoxy led to the exclusion of dissidents like Pailthorpe and Mednikoff
(and Agar and Ithell Colquhoun) from the London Surrealist Group. Grace
Pailthorpe continued on her own way, as her painting *The Spotted Thrush*,
1942, shows, still striving to reunite conscious and unconscious in her depiction of the two traumas of birth and (maternal) sustenance.

Ithell Colquhoun

Ithell Colquhoun, 1906–1988, said of herself in 1939: 'I learned to draw at
the Slade School. I have not yet learned to paint. Sometimes I copy nature,
sometimes imagination . . . My life is uneventful but I sometimes have an
interesting dream'. Originally she had come into contact with Surrealism
through Dali's paintings when she was living in Paris in 1931. Colquhoun practised both automatic writing and automatic drawing and she experimented
with collage. Throughout the 1930s and 1940s she was also working in her
own specific field, on what Herbert Read called 'psychomorphology'. Like
Pailthorpe, Colquhoun was a theoretician of surrealist practice, and defined
psychomorphology as 'an effort to tap that level of consciousness sometimes
perceptible between sleeping and waking, which consists in coloured organic,
non-geometric forms in a state of flux'. That artistic principle produced
distorted, biomorphic elements modified by natural agents, or 'interpreted' –
in other words misused or misnamed objects. This classification had been the
structure of the Surrealist Objects exhibition which took place in the London
Gallery in 1937. Colquhoun's time with the Surrealist Group in London was
intense but short. She joined the Group in 1939, was married briefly to one
of its leading figures, Tonio del Renzio, wrote for its publications but was
finally excluded by Mesens in 1940. Her mistake was not to try to marry
Surrealism and psycho-medical science like Grace Pailthorpe but to delve too
deeply into the occult. Reproductions of her disturbing paintings can be seen
in Sidra Stich's *Anxious Visions*. After leaving the Surrealists, Ithell Colquhoun
turned to Rosicrucianism.

Edith Rimmington

Edith Rimmington, 1902–1986, was introduced to E.L.T. Mesens and the
Surrealists by Gordon Onslow-Ford in 1939, the same year as Ithell Colquhoun

with whom she practised automatic drawing and writing. Her work between 1939 and 1944 consisted mainly of her contribution to the periodicals of the London Group. Her drawings and collages were reproduced in *Arson* in 1942, and in *Fulcrum* and *Message from Nowhere* in 1944. *Arson*, published only once in March 1942, has been described as 'an enthusiastic attempt to give a clear focus on Surrealism', to reinject a new energy in the Group in those days of a too evidently individualist policy. *Arson* was a Manifesto, signed by Conroy Maddox (member of the Birmingham Surrealist group) as well as by John Melville, Edith Rimmington, Emmy Bridgwater, Robert Melville, Eileen Agar, and Tonio del Renzio. It also contained reproductions of Agar and Bridgwater's works.

L'Oneiroscopist (a Franco-English title), 1942, is one of Rimmington's most striking paintings. Its interpretation epitomizes most of the artist's concerns. A huge bird-like creature reminds us of Max Ernst's obsession with all that flies. This long-beaked, yet vaguely human animal is wearing a scaphander (cork swimming belt) .

The Oneiroscopist literally means: the specialist in looking through dreams. This, of course, is a reference to the philosophy of Surrealism; it is at once a technique of introspection, of self-knowledge, and a method for understanding dreams. The diver represented here is the artist swimming down to the marine depths of the unconscious. The submarine world is indeed an important element in the women Surrealists' imagery. We feel almost familiar with this peculiar animal, the impersonation of the Surrealist creator; like the bird, she can go higher and deeper into her own mind. Of course – but that is only the logic of the absurd – the bird cannot fly, because it is human. All ideas and images come from the 'interior model', in opposition to figurative art, which gives the priority to perceptions of the external. Interiority is the submerged new continent yet to be explored; the artist is the medium between the shores of the river of Forgetfulness – hence the title of another Rimmington painting, *Washed in Lethe*.

Emmy Bridgwater

Emmy Bridgwater, 1906– , like her close friend Edith Rimmington, experienced the International Surrealist Exhibition in London in 1936 as a revelatory 'transformation', which freed her to take a fresh approach to her own work. That work, consisting of paintings, drawings, and poems, reveals a more spontaneous, almost natural perception of Surrealism. Automatism meant for Bridgwater 'following what my hand does'. She enjoyed it as an entertaining way of expressing new ideas and feelings. Her inspiration contrasts with the more violent visions of some of her friends in the London Group.

Beware of the View is an undated ink drawing. It consists of an unbroken line, equivalent for a single player to the Surrealist game called 'exquisite corpse' the final goal of which is a collective drawing, each participant completing what is left for him or her to see. The line links disparate, formless objects within a definite frame or 'view'. This view is being watched through a telescope by a man, his legs held together in a triangle. It could be a humorous criticism of the 'serious nonsense' the Surrealists were reproached for or a criticism of Surrealism's critics. All Bridgwater's ink drawings, made between 1940 and 1948 seem to be inspired by Edward Lear's Nonsense Book.

When Emmy Bridgwater started exhibiting her work in the 1940s, her concern for the world situation and her reflections on the war revealed a more humane and more clear-cut conception of the Surrealist task than had prevailed hitherto. It includes the lines:

> Black death and watered down trees crying
> Out shrieking with 'it is time,
> Now it is time,
> And soon there will be no time'.
> (See Michel Remy's Introduction to Mayor Gallery Exhibition
> Catalogue, *British Surrealism: Fifty Years On, 1986.*)

Bridgwater's paintings *Remote Cause of Infinite Strife*, reproduced in *Arson* in 1942, and *Tomorrow's Choice* point to the artist's consciousness of something needing to be done and of her own limited power. From a technical angle, they are like collages of unrelated elements, mixing a portrait with a landscape for example. The absence of perspective adds to the impression of a non-professional, child-like spontaneity. They testify to the artist's freedom from intellectual pretence, which could explain why her acknowledgment within the Group was not easy, and why she may have felt like an outsider in some ways: she let her works talk for themselves.

The 'Living art in England' Exhibition organized in support of the Czech and Jewish refugees in January 1939 showed among other powerful paintings Gwenedd Reavey's *Window on Europe*. This still completely unknown artist could represent the pessimistic side of Surrealism, if such an expression was not contradictory. Hands stretching out of the fractured concrete are waving for help, while in the background a tower-like woman is collapsing in an expression of sheer despair. Reproduced in the catalogue of the exhibition in the *London Bulletin*, this painting anticipates the sorrow and horror of Bridgwater's poem. Pacifism and its artistic translation had a vitally important part to play in Surrealism, although it did not necessarily match with the disinterested functioning of the mind, and with the deliberately – too deliberately? – ludic activities of the male artists. The English Group's official position against war, and that of the Parisian Surrealist group, were almost exclusively expressed through manifestos, leaflets, or pamphlets. In contrast, the women, as we saw in Lee Miller's photographs, were able to express their anger in a directly creative way. This could be one explanation of the fact that the London Group's manifestos were very seldom signed by its female members.

Conclusion

What brought all these women artists into the Surrealist Movement was their shared passion for experimenting in visual art along with the need they recognized for loose, liberating adherence to a community of daringly creative people in order to produce and show what would otherwise be rejected as 'impossible'. Another ingredient in their community of thought was humour, for humour rather than any academic art qualification was the essential password permitting participation among the Surrealists. Although surrealist humour may sometimes seem inadequate, purposeless or even strained, it was still a vital attempt to negate all that suppresses our ability to laugh – even desperately laugh – at 'normal' life. Seriousness was held to be counterproductive, insofar as it subverted the vital role of dreams and suppressed the boundless flow of a totally uninhibited imagination.

Nevertheless, seriousness, like cheerfulness, has a habit of breaking in – all too understandably in an age of Fascism and World War. And although the women artists in Britain never formulated a joint counter-policy or constituted a collective counter-force to the dominant male Surrealists, they did produce in their very diverse work what amounted to a counter-culture, one in which seriousness was not altogether outlawed. Instead of their male counterparts' indulgent fantasies of 'Mad Love' and 'Eternally Youthful Faceless Beauty', they resorted to asexual imagery, distancing themselves from the risk of mere objectification as women and creating instead a kaleidoscopic image of women also alienated, but alienated with a difference. As Eileen Agar said, they felt themselves to be a minority within a minority, taking 'deviance as a principle of creativity'.

All the women experienced loss, either in madness, broken relationships, or war; yet they tried to find a new unity through art, working with fragments of (transmuted) nature or rejected pieces of civilization. They also focused on the positively creative aspect of femaleness. However diverse the work of the women artists was and however much they deviated from some of the unorthodox orthodoxies of the men (notably when Agar, Pailthorpe, and Colquhoun all chose autonomy rather than domination by Mesens in 1940) all Surrealists shared a fascination with the juxtaposition of clashing elements in their visions of an inner reality other than that of unquestioning 'common sense'. And the women artists, just as much as the men, *played* with the fundamental contradiction of art, its attempt to make something wonderful out of a frightening world.

References

AGAR, E. (1988) *A Look at My Life*, London, Methuen.
BILLETER, E. and MICHETTE, C. (Eds) (1987) *La Femme et le Surréalisme*, Lausanne, Catalogue Musée cantonal des Beaux Arts.
BRODERIE, M. (Ed.) (1977) 'La Femme Surréaliste', *Obliques*, no. 14–15, Paris.
CARRINGTON, L. (1989) *The House of Fear*, London, Virago.
CAWS, M., KUENZLI, R. and RAABEY, G. (Eds) (1991) *Surrealism and Women*, Cambridge, Mass., MIT Press.
CHADWICK, W. (1985) *Women Artists in the Surrealist Movement*, London, Thames and Hudson.
DEEPWELL, K. (1992) *Ten Decades: Careers of Ten Women Artists born 1897–1906*, Norwich City Gallery Catalogue.
GRIMES, T., COLLINS, J. and BADDELEY, O. (1989) *Five Women Painters*, Oxford, Lennard.

Brigitte Libmann

LONDON BULLETIN, 1938–1940.

PENROSE, R. (1985) *The Lives of Lee Miller*, London, Thames and Hudson.

PENROSE, R. (1981) *Scrapbook 1900–1981*, London, Thames and Hudson.

RAY, P. (1971) *The Surrealist Movement in England*, New York, Cornell University Press.

READ, H. (1936) *Surrealism*, London, Faber.

ROBERTSON, A., *et al.* (1986) *Surrealism in Britain in the Thirties*, Leeds City Art Galleries.

SCHLIEKER, A. (1991) *Leonora Carrington: Paintings, Drawings and Sculptures 1940–1990*, Serpentine Gallery Catalogue.

STICH, S. (1990) *Anxious Visions*, New York, Abbeville Press.

Chapter 15

Hilda Matheson and the BBC, 1926–1940

Fred Hunter

One of the problems faced in studying the pioneer women in broadcasting is the very ephemeral nature of their product, irretrievably lost in the ether when recording was still virtually unknown. How many people know, for example, that it was a woman who helped develop the concept of 'the scripted talk', which brought to a potential audience of five million British 'listeners-in' the voices of the outstanding writers of the 1930s, as well as commissioning Harold Nicolson to discuss new styles in literature and enabling him to play a record of James Joyce reading from *A Work in Progress*? That woman was Hilda Matheson, who today only appears in books about other people's lives where she features, usually, in a lesbian relationship with someone better known than herself.[1]

Born on 8 June 1888 in Putney where her father was a Presbyterian minister, as a teenager she became fluent in French, German, and Italian when her father's breakdown in health necessitated living abroad. In 1908, with his health restored and living in Oxford, his daughter enrolled in the Society of Oxford Home-Students (now known as St Anne's College), where she studied history. Writing about her time at the college, Matheson recalled that 'we [felt] in the very van of progress. I suspect that each generation of women students has felt very much the same'.[2] Even so, to those Home-Students in the first decade of this century, many of whom, like Matheson, lived at home, 'the University, on the side of its undergraduate activities, seemed to us marvellous and remote [but] one had little dealings with them, whatever one might do unofficially'.[3]

Leaving college in 1911, Matheson's first job was as a part-time secretary to H.A.L. Fisher, at New College, before working under the Keeper of the Ashmolean Museum, David George Hogarth, in whose presence she met T.E. Lawrence ('Lawrence of Arabia'), just after the fall of Damascus, in 1918. During World War One she was employed, as were so many more women, to work in army intelligence at the War Office: 'deep in MI5' as her mother described it in her memoir.[4] Later she was sent to the British Mission in Rome with the task 'of forming a proper office on the model of MI5 in London'.[5] When the war ended, in 1919, Matheson, after initially turning down the offer, became political secretary to the first woman Member of Parliament, the American-born Nancy Astor. This put Matheson centre-stage in the worlds of politics, letters, and society. Her competence and the fact that she 'knew everybody' so impressed John Reith, when she visited him on Lady Astor's behalf in 1926, that he persuaded Lady Astor to release her to work for him at the fledgling British Broadcasting Company. In thanking Lady Astor for the farewell gift of a cheque, Matheson wrote:

> I was already feeling ready to howl with misery at your kindness . . .
> [and now] . . . your reckless generosity . . . and thoughtfulness in
> devising that way of making me an independent capitalist . . . I hate
> going, as I have seldom hated anything, and I have loved all the time
> I have been with you.[6]

What was it that attracted her away from Lady Astor? The offer of a position to influence the way a new medium of communication might be developed was certainly not one she could refuse. Immensely energetic, she 'knew everyone' at a time when Reith, whose primary job was the organization, was out of touch with the literary and educational world. Nominally employed to assist J.C. Stobart administer the BBC's Education Department, but really to launch a new department, Matheson did not believe, like Vita Sackville-West, that 'woman *cannot* combine careers with normal life'.[7] Hilda Matheson was soon embroiled in the debate, sparked by the General Strike in May 1926, about the BBC's role as a news provider. During the strike the BBC had, for the first time, used its own reporters to collect news and attempted to present news bulletins 'detached from both sides of the conflict'.[8] However, the Labour MP Ellen Wilkinson declaimed that *she* felt like 'asking the Postmaster General for my licence fee back',[9] so biased did *she* consider the BBC coverage. Reith supported the government because he believed it was acting in the national interest against disruptive sectional interests.

After 1927, when it became the British Broadcasting Corporation, the BBC was permitted to broadcast its first bulletin at 6.30 p.m. Publishers of evening newspapers had been against earlier bulletins, regarding them as unfair competition. Each year the BBC could also broadcast 400 eye-witness accounts [i.e. news reporting] and running commentaries on sporting events. Yet the BBC still received its bulletins *written* by the Reuters news agency, though they were collated by BBC staff, and it was not until 1929 that Reuters and the Press Association agreed to supply their full wire services to the BBC, for its staff to select and sub-edit for bulletins. Since this meant that the BBC could then begin to write its *own* news bulletins, the immediate consequence was the establishment of a small News Section, in 1927, under Hilda Matheson as 'Head of Talks'.[10] Thus, by the end of the 1920s, the BBC had won the right to provide the new listening public with news and debate on the major issues of the day. And here Matheson was something of an innovator, securing the co-operation of government departments in providing their information direct to the BBC. Soon the BBC news bulletins were cluttered with these releases, so much so that, by 1931, most of it was relegated to a ten-minute programme on Thursday evenings.

In establishing the BBC's first News Section, Matheson sought advice from the Royal Institute of International Affairs, one of whose founders was

Nancy Astor's husband, Waldorf, Viscount Astor, and from Geoffrey Dawson, editor of *The Times*, to help her assess what problems the BBC might face by becoming a major provider of news. On their advice she commissioned a former assistant editor of the *Westminster Gazette*, Philip Macer-Wright, to work in the News Section during the summer of 1928 and then to present his observations and recommendations based on his experiences.[11]

Macer-Wright's eleven-page report, essentially the most important document on news values ever produced for the BBC, was constantly referred to during the ensuing decades as the BBC kept refining its news presentation policy. He described in detail the machinery needed for a fully-fledged news room, and outlined how senior staff would need access to 'accredited experts' on financial, sporting, legal, and scientific matters. Macer-Wright's recommendations not only confirmed existing trends within the News Service, but also stiffened management's resolve to wrest control of the news from the news agencies, who were reluctant to acknowledge a new medium of communication, which they saw as threatening to them and to the newspaper press. Macer-Wright asserted that if the BBC wanted to make the news service attractive to the millions of listeners, it could not afford to ignore the appeal of human interest news, simply and attractively conveyed. Radio news, he said, must have 'news values'[12] and he also wanted bulletins arranged in fixed categories with Home News coming first, followed consecutively by Overseas News and Sports News. He also recommended the presentation of a bulletin especially written for listeners' ears. This was something that Matheson, and the Talks Department, had pioneered and to which she had, personally, devoted much time, training would-be broadcasters how to write for the ear. By comparison, the Reuters bulletins were written in involved, cumbersome, and florid *print* journalese, most of which had to be rewritten to make it suitable for reading aloud on the wireless. One of her colleagues credited Matheson with 'discovering, by trial and error, the "technique" of the spoken word over the air'.[13]

Scannell and Cardiff note that the Talks Department under Matheson, from 1927 to 1932, and her successor Charles Siepmann, from 1932 to 1935,

> was inspired by a common commitment to the importance of radio as a new form of social communication, and a common interest in developing effective methods of communicating via the spoken word. Matheson was a woman of courage, originality and culture, and she brought these qualities to broadcast talks.[14]

Those who worked with her described Matheson as 'enterprising, indefatigable, and liberal-minded [with] a sympathetic personality, capable of winning and holding the loyalty of subordinates'.[15] R.S. Lambert later recalled her as 'toiling single-mindedly, night and day [making] the Talks Department a live, energetic and humane department'[16] of the BBC. In Lambert's view,

> Hilda Matheson's outlook was that of the typical post-War Liberal, with its idealistic internationalism expressed in mistaken devotion to the League of Nations, its sympathy with Socialistic experiment, its cultivation of the innovating schools of poetry and art, its enthusiasm

for feminism. She brought to the microphone – often persuading with her sympathetic tongue those who had hitherto been hostile or contemptuous – many of the most important broadcasters of our time . . . Wells, Shaw, Harold Nicolson, Winston Churchill and Lady Astor.[17]

Matheson also introduced Vita Sackville-West to the perils of broadcasting, and the two had an intense personal relationship which lasted from 1928 until early 1931, when Matheson was replaced in Vita's affections by another Oxford graduate, Evelyn Irons, of Somerville College, who was women's page editor of the *Daily Mail*. Vita Sackville-West has vividly described what it was like to broadcast a talk on the wireless when she wrote:

> You are taken into your studio, which is a large and luxuriously appointed room, and there is a desk, heavily padded, and over it hangs a little white box, suspended from two wires from the ceiling. There are lots of menacing notices about: 'DON'T COUGH – you will deafen millions of people', 'DON'T RUSTLE YOUR PAPERS', and 'Don't turn to the announcer and say "was that all right" when you've finished'. One has never talked to so few people and so many; it's very queer.[18]

Matheson even had Vita and her husband, Harold Nicolson, who both preferred sexual relations with people of their own sex, discussing marriage and, as Nicolson remarked to Matheson, 'We won't be able to mention sex, I presume'.[19]

According to Lambert, Matheson was invariably tactful and persuasive in her defence of the line of action she believed to be correct and outstandingly successful in the production of ideas, the planning of programmes, and in contact with speakers; but she made enemies by the very persistence of her memoranda, and by the way she sought to extend the influence of the Talks Department in all directions. Someone who worked closely with her at the BBC, Lionel Fielden, felt that both Reith and Matheson

> were the victims of circumstance. Voices whispered to him that he was being RUN by a gang of REDS; he made dictatorial gestures; you took up a cudgel; he became domineering, you wild; until at last there was nothing for it but your resignation. But the real cause, I feel, lay in the twin spectres of Hate and Fear, which in 1930, were creeping back on to the world stage. The Blimps were on the warpath and you and your kind were doomed.[20]

For another of her BBC colleagues, Matheson's departure from the BBC was the result of one of the many 'misunderstandings' that chequered the first decade of the BBC's history. But to Reith, who had 'developed a great dislike of Miss Matheson and her works' [his diary contains references to 'the Red Woman'] it came as a relief and 'to my much embarrassment I had to hand over the staff's present to Miss Matheson' when she left the BBC in January 1932.[21]

To the *New Statesman and Nation* her departure from the BBC was a disastrous turning point. Matheson chose to resign because Reith refused to allow Harold Nicolson to praise *Ulysses* in a projected talk. But she realized that Reith wanted a return to less controversial talks and, in a draft resignation letter, she wrote that she 'could not loyally administer a policy which seemed to be turning into a reversal of what I had been instrumental in helping to build up'.[22]

Always Matheson's stalwart supporter, Lady Astor suggested that Matheson be appointed to the BBC Board of Governors![23] However, with Siepmann in charge of Talks, Matheson's ideals still flourished until his departure, in 1935, when there was a definite move away from her, liberal, ideas. But her friends were still aware of her liberal ideas and it was with regret that Matheson turned down an offer from Leonard Woolf, in 1932, to run The Hogarth Press.[24] Matheson had met his wife, Virginia Woolf, through her friendship with Vita Sackville-West.

Throughout the 1930s Matheson continued to combine journalism and publishing, working as radio critic on the Astor-owned *Observer*, and as a weekly columnist in *The Week-End Review*, as well as publishing a book on broadcasting in 1933. The decision, by H.A.L. Fisher, to commission this book for the Home University Library, Lord Reith described as 'monstrous',[25] without realizing that Matheson's first job had been as a part-time secretary to Fisher, in Oxford, when he had originally set up the *Library*. Between 1933 and 1935 Matheson worked at the Royal Institute of International Affairs (Chatham House), a major interest of Lord Astor, on the *African Survey* (1938), for which she was awarded the OBE. Its nominal author, Lord Hailey, who was taken ill at a vital stage in the research, commented that 'but for her initiative and determination it might never have seen the light'.[26] Indeed, one obituarist noted that 'she took on a great deal of extra work [as] secretary for the enterprise, brilliantly, in a way that would have been quite beyond most men's powers'.[27] Her friend, Dame Ethel Smyth, commented that her greatest 'fault was her inability to say "No" when asked to do a service', adding that she blended 'intellectual grip with ... perfect manners of soul'.[28] In her book, *Broadcasting*, Matheson commented on how demanding work was in those early days, when programmes could *not* be pre-recorded and those 'who worked all day in the office may be in the studio all evening directing a programme'.[29]

With the World War Two looming, Matheson again found herself involved in broadcasting pro-British propaganda as Director of the Joint Broadcasting Committee, which purported to promote 'international understanding by means of broadcasting'. In July 1939, she wrote to F.W. Ogilvie, the new Director General of the BBC, that the JBC would have greater freedom than the BBC to develop propaganda for the German audience in the critical months of August and September of that year.[30]

With her friend and lover, Dorothy Wellesley, she was also involved in another government-subsidized publishing venture, designed to counteract German propaganda abroad, called *Britain in Pictures*.[31] But, by the end of October 1940, Hilda Matheson was dead and the BBC eventually took over the running of the Joint Broadcasting Committee in July 1941.[32]

Her early death, at 52, was keenly felt, and Dorothy Wellesley, the seventh

Duchess of Wellington, erected a plaque to her memory in the grounds of Penns in the Rocks, at Withyham in Sussex, which reads simply 'Amica Amicorum' – 'friend of friends'.

Notes

1 See Victoria Glendinning (1984), *Vita: The Life of Vita Sackville-West*, Harmondsworth, Penguin Books.
2 R.F. Butler and M.H. Pritchard (Eds) (n.d.), *The Society of Oxford Home-Students: Retrospects and Recollections (1879–1921)*, Oxford, p. 113.
3 *Ibid.*, p. 114.
4 [Mrs Meta Matheson] (1941), *Hilda Matheson*, Letchworth, The Hogarth Press. Internal evidence in the Hogarth Press Archives at the University of Reading suggest this was seen through the press by V. Sackville-West (although she considered the contents not equal to the stature of Hilda Matheson). Contributors included Philip Noel-Baker, MP, and Vernon Bartlett, MP, H.G. Wells, Mrs H.A.L. Fisher, Lady Astor, and others.
5 *Ibid.*, p. 10.
6 Lady Astor Archive, University of Reading, MS 1416/1/2/37.
7 Glendinning, *Vita*, p. 209.
8 Paddy Scannell and David Cardiff (1992), *A Social History of British Broadcasting. Volume One: 1922–1939*, Oxford, Blackwell, p. 33. Unlike other histories of broadcasting, this book redresses the balance in Matheson's favour, although the authors are incorrect in stating that she served for several years as Vita Sackville-West's secretary when she left the BBC.
9 *Ibid.*, p. 31.
10 *Ibid.*, p. 41.
11 *Ibid.*, p. 113. Macer-Wright's report is in the BBC Written Archives Centre (WAC), R28/177/1: 'Suggestions for the improvement of the BBC News Service', 24 September 1928.
12 *Ibid.*, p. 114.
13 [Matheson], p. 36, and Scannell and Cardiff (1992), pp. 161ff.
14 *Ibid.*, p. 153.
15 R.S. Lambert (1940), *Ariel and All His Quality*, London, Gollancz, p. 62.
16 *Ibid.*, p. 63.
17 *Ibid.*, p. 64.
18 Glendinning, *Vita*, p. 193.
19 *Ibid.*, p. 214.
20 [Matheson], p. 36.
21 Lord Reith's MSS *Diary 1930–33,* December 1930 and January 1932, BBC WAC.
22 Matheson to Reith, undated draft letter in Astor Archive, MSS 1416/1/1/962.
23 *Ibid.*, letter to Major W. Elliot, MP, 9 December 1931.
24 Leonard Woolf Papers, University of Sussex, Sx MS 13.
25 Reith, *Diary*, January 1932.
26 [Matheson], p. 45.
27 *The Times,* 7 November 1940.
28 *The Times*, 6 November 1940.
29 Hilda Matheson (1933), *Broadcasting*, London, Home University Library, p. 57.
30 Asa Briggs (1970), *The History of Broadcasting in the United Kingdom. Volume III: The War of Words*, London, Oxford University Press, p. 185.
31 Matheson to Astor, 14 October, 1940, Astor Archive: 'I have been engaged since before the war in propaganda, of various kinds, abroad'.
32 Briggs (1970), p. 344.

Chapter 16

'Nothing is Impracticable for a Single, Middle-Aged Woman with an Income of her Own': The Spinster in Women's Fiction of the 1920s

Maroula Joannou

In Katherine Mansfield's short story, 'Miss Brill' (1922), an ageing spinster assumes a practised air of *joie de vivre*. The shabby fox-fur fondly paraded by Miss Brill in all weathers is a totem belying the reality of its owner's poverty and loneliness. The truth is brought home by a courting couple overheard in the park. The boy resents the 'silly old mug' of the 'stupid old thing' and the girl callously mocks the fur that to her looks 'exactly like a fried whiting'. The story ends pointedly. Miss Brill carefully returns her cherished fur to its box, 'but when she put the lid on she thought she heard something crying'.[1]

The disparity between how the spinster sees herself and how she seems to others is the essence of this short story, and of much imaginative writing on spinsterhood published in the 1920s. Women writers had long been concerned with the plight of the spinster, a concern reaching back to Charlotte Brontë's *Shirley* (1851) and resurfacing with Mona Caird's *The Morality of Marriage* (1897), Cicely Hamilton's *Marriage as a Trade* (1909), and Elizabeth Robins' *Way Stations* (1913), as well as F.M. Mayor's *The Third Miss Symons* of the same year. But the topic of spinsterhood recurs in the 1920s with a frequency and insistence that can only be understood in its historical specificity. With unmarried women of marriageable age far outnumbering men of the same age group the meaning of spinsterhood became a site of contestation between those who wished to objectify the spinster and others who saw her as a person with needs, desires, and potential of her own.

If much that was written in the 1920s was unashamedly hostile to the spinster, there was also a strong drive towards self-assertion on the part of spinsters, to which the texts to which I shall refer bear powerful witness. But the writings of Ivy Compton-Burnett, Radclyffe Hall, Winifred Holtby, Katherine Mansfield, F.M. Mayor, Dorothy Richardson, May Sinclair, and Sylvia Townsend-Warner are diverse, and cannot, without gross distortion, be assimilated to any common label. There was not, nor could there have been, sufficient consensus, aesthetic or political, among women writers to make a uniform approach to spinsterhood possible. Nevertheless, common interests, for example in the wider options that should be open to the increasing numbers of women whose minds did not travel along the smooth road to matrimony, are explored across a wide spectrum of literary texts. And a number of oppositions are frequently invoked: movement or stasis? duty to self or to others? freedom or imprisonment?

For most of the inter-war period about a third of all women who had not married by the age of 29 did not marry during their reproductive years.[2] As Billie Melman has pointed out, social observers of the day frequently made use of a corrupted Darwinist vocabulary, and the socio-demographic imbalance was widely 'interpreted as a sign of universal disequilibrium – a fall, as it were, from a "natural" state of harmony between males and females'.[3] After the Census disclosures in the late summer of 1921 speculation on the future prospects of two million 'superfluous women' became a national pastime. The *New Statesman* highlighted the difference between its own attitudes and those of other publications in this unsigned editorial piece on 24 September:

> Hence we think it absurd on the part of the Press to talk so much of the two thousand [*sic*] 'superfluous women', or 'thwarted women', as though 'superfluous', 'thwarted', and 'unmarried' were convertible terms. . . . We have known a considerable number of married women who were at least as superfluous as St. Teresa, and there are married men of our acquaintance who live thwarted lives in comparison with William Pitt. Who, indeed, is to decide what is a thwarted life? Was Swift's a thwarted life? Or Ruskin's? If so, it seems clear that thwarting may be merely a means to completer fulfilment.[4]

The tone in which the public debate was conducted was not to every woman writer's taste. Some, including May Sinclair, whose novels *Mary Olivier: A Life* (1919) and *Life and Death of Harriett Frean* (1922) are concerned with the effects of sublimation and repression on the life of the spinster, pointedly refused to participate. When Sinclair, who was living in genteel poverty with her mother at the time, was asked to write for the newspapers she answered: 'I'm not interested in the Superfluous 2,000,000! If they *are* superfluous, let them emigrate. But they don't and won't'.[5]

The notion of the state-directed emigration of women to the colonies was perhaps the most authoritarian of many 'solutions' to the demographic imbalance. But when *The Times* of August 1921 advised women to travel to the colonies to find a husband, as Adela Quested does in E.M. Forster's *A Passage to India* (1924), Vera Brittain wrote to Winifred Holtby ('one "superfluous" woman to another') making her feelings clear:

> Personally I haven't the least objection to being superfluous . . . though I shall be delighted for any work I may do to take me abroad, it will not be because I shall thereby be enabled the better to capture the elusive male.[6]

However, in practice the life of the spinster in the family home was still often one of unrelieved domestic servitude. Radclyffe Hall, visiting a Devon resort in 1921, was distressed by the sight of an old lady waited upon by her ageing daughter: 'ghastly to see these unmarried daughters who are just unpaid servants', and the old people 'sucking the life out of them like octopi'.[7] Hall later discarded her first title for *The Unlit Lamp*, 'Octopi', but retained the metaphor of strangulation from which the novel's social satire derives much of its strength. The inter-war novels of Ivy Compton-Burnett, who had experienced psychic subjugation at the hands of her widowed mother, are peopled with insupportable autocrats of the breakfast table. Domestic tyranny exercised over their young by matriarchs – the egocentric, Sophia Stace in *Brothers and Sisters* (1929), the sharp-tongued Harriet Haslam in *Men and Wives* (1931) – is a commonplace.

The loosening in the 1920s of sexual taboos extended only to non-marital heterosexual activity.[8] It was of little benefit to the many unmarried women with neither the desire nor the opportunity to form sexual relationships with men, or whose sexual orientation was to women. Indeed, the revolutionary feminist historian, Sheila Jeffreys, has claimed that the hostile attitudes to the spinster were deliberately aggravated by sex reformers who stressed women's right to experience sexual pleasure in marriage, reviling earlier feminists as prudes and puritans.[9] This claim should, however, be viewed with caution.[10]

The taboos on homosexual activity, exemplified in the trial of *The Well of Loneliness*, intensified considerably during the decade, doing much to break the patterns of emotional dependency between women in the 1914 war. Moreover, a woman attempting to escape from the traditional limitations of domestic obligations might well find herself thrust into them again by the authority of the psychologist. Winifred Holtby noted with dismay that women everywhere were being told to 'enjoy the full cycle of sex-experience, or they would become riddled with complexes like a rotting fruit'.[11]

In particular, there was a vogue for psychoanalytic ideas after more of Freud's works, available since 1913, came to be translated in the 1920s. Katherine Mansfield rebelled openly against the sexualization of literary sensibility: 'I shall *never* see sex in trees, sex in the running brooks, sex in stones and sex in everything'.[12] But for many women writers the confusion and conflict stirred by the cross-set of taboos and new modes of analysis was deeply distressing, and, as Winifred Holtby commented 'the wonder is that any woman continued to write novels at all'.[13]

Hostility to the spinster was at its most vitriolic in the Catholic social theorist, A.M. Ludovici. Ludovici combined his onslaughts upon 'thousands of bitter or sub-normal women' who had 'thwarted or deficient passions'[14] with contradictory attacks on 'Feminist propaganda in favour of sterile Free Love, Restricted Fertility and Birth Control'.[15] In *Woman: A Vindication* he had uttered this stern warning:

Since the spinsters of any country represent a body of human beings who are not leading natural lives, and whose fundamental instincts are able to find no normal expression or satisfaction, it follows ... that the influence of this body of spinsters on the life of the nation to which they belong, must be abnormal, and therefore contrary to the normal needs and the natural development of that nation.[16]

'Natural Lives', 'normal expression', 'abnormal' influence and 'natural development' – the play on the idea of 'the natural' in this litany of deviance reveals an obsessive interest in exorcising difference. In a labyrinth of connection, interrelation, and extension, the spinster is transformed into an agent of perverse subversion, a fifth columnist, an enemy of the state. Ludovici's views on spinsterhood are clearly extreme, but even so repay careful attention as symptomatic of the opinions expressed commonly, if less virulently, at the time. As late as 1935 Dorothy Sayers' detective, Harriet Vane, investigating a spate of poison-pen letters in an Oxford college, 'could think of whole sets of epithets, ready-minted for circulation' for their author, among them ' "soured virginity" . . . "unnatural life" . . . "semi-demented spinsters" . . . "starved appetites and suppressed impulses" '.[17]

But ironically those like Ludovici who labelled the spinster deviant and malignant conferred an authoritative right to speak upon the woman thus travestied. This accelerated the process whereby the derogatory image of the spinster came to be contested, and in the end transformed, largely through the objections voiced by spinsters themselves. The spinster, according to Ludovici, is defined by 'want' or 'lack'. The spinster must have some 'wants' but is denied others. The spinster must 'want' a husband. This 'want' is confidently refuted in *Lolly Willowes*, in Winifred Holtby's *The Crowded Street*, in *The Unlit Lamp*, in *Surplus*, and in Sheila Kaye-Smith's *Joanna Godden*. It is beautifully subverted by E.H. Young's *soi-disant* spinster, Hannah Mole in *Miss Mole* (1931), jauntily sailing under spinsterhood as a flag of convenience, and wondering why so much importance was attached to the chastity of women.

Although there is always a ricochet between lived experience and literary experience, the literary and lived experience of women in the 1920s reveals a particularly intense interchange. In *Time and Tide*, the journalist Rosaline Masson characterized the plight which commonly befell the dispossessed daughter in an arresting image:

> Much of the home-life of our country is built up on a substratum of obscure martyrdoms . . . until the family home is broken up by death. And then the luxurious, hospitable family home belches forth dismayed spinsters into an unsympathetic world, to wander about as aimlessly, and seek cover as nervously, as do the wood-lice when the flower-pot is lifted.[18]

Sylvia Townsend Warner's heroine, Lolly Willowes, is 28 and must 'make haste if she were going to find a husband before she was thirty'.[19] When her father dies, a bemused Lolly, 'feeling rather as if she were a piece of family property forgotten in the will, was ready to be disposed of' (p. 6) as her relations should think best. Like Lady Slane in Vita Sackville-West's *All Passion Spent* (1931) Lolly believes it 'best as one grows older to strip oneself of possessions, to shed oneself downwards like a tree, to be almost wholly earth before one dies' (p. 106). When she wishes to leave his London town house for the countryside, brother Henry is aghast. But kneeling thankfully among the cowslips, 'the weight of all her unhappy years' behind her at last,

Lolly trembles with relief, 'understanding for the first time how miserable she had been' (p. 149) before her release from the wretchedness of dependence.

The myths that passed for truth about the spinster's life were not merely journalistic excesses, nor simply the currency of notorious misogynists. On the contrary, they were so resilient and common that it was all but impossible for the spinster to escape the socially imposed expectations of the dominant sex. The spectre of loneliness, the chimaera of romantic love, the reality of interminable waiting, for a husband who never quite materializes on the threshold, and for a life that consequently never quite begins, all these also found resonances in women's fiction. The symptoms of women forced to repress deep emotional and sexual needs were vividly evoked by May Sinclair. The sickness of Aunt Charlotte in *Mary Olivier* is clearly related to her un-requited longing for marriage. Harriett Frean's lingering illness is the learned helplessness of one who thinks of herself as a girl long after becoming a woman. In later years if Harriett were introduced to any stranger 'she accounted for herself arrogantly: "My father was Hilton Frean"'.[20] The red campion outside Harriett's house symbolizes the passion forbidden her, the brittle blue ornamental egg, a wedding present to her parents, the fertility she is never to know. In quiet desperation Harriett clings to the image of her lost mother, 'and always beside it, shadowy and pathetic, she discerned the image of her lost self'.[21] The mental disintegration that gradually overtakes Harriett nearly befalls Mary Olivier, who cannot marry because her mother 'needs her' but just escapes Harriett's unhappy fate through sublimation in philosophy and poetry.

Yet women writers did not all create spinster protagonists content to submit meekly to their lot. Some chose to become actively engaged in a protracted struggle for self-realization. People might speak pityingly of the nation's spinsters: 'nothing for them except subjection and plaiting their hair' (*Lolly Willowes*, p. 235). But many a spinster was confident that the last thing she required was pity. And a succession of sympathetic works all reject the prevalent condescension and contest the simple caricatures of the man-hating old maid and the old maid desperate to find a man.

As women's literary resistance to patriarchal values took subtle and diverse forms, the presence of a 'strong' or overtly rebellious character is clearly inadequate as a yardstick of feminist consciousness. It is not, after all, Mansfield's timid and ineffectual spinster sisters in 'The Daughters of the Late Colonel' (1922) who exemplify resistance to the crushing weight of patriarchal authority, but the author's sensitive atomising of the sisters' predicament which makes possible the story's incisive analysis of patriarchy. The cumulative power of the narration, symbolism, and imagery pinpoints the 'sunlessness' of Josephine and Constantia's drab existence, and intimates that freedom, like the reprieve that arrives after the executioner's axe has fallen, has come too late to be of use. The more resolute sister, Constantia, is allowed one brief act of self-assertion: locking the drawer upon her father's belongings (at a symbolic level locking her father's lingering presence out of her life). Constantia's gesture of defiance is uncharacteristic, unexpected. It does, however, illustrate that routine submission to authority need not exclude the possibility of occasional resistance, any more than discomfort with patriarchal values need imply their complete or irreversible rejection on a woman's part.

The lives of dependent unmarried women were not merely restricted, they were potentially shot through with pain. To track the life of the spinster from childhood was to choose to explore the reasons why so many women beyond the usual age of marriage, 'growing old, as common as blackberries, and as unregarded' (*Lolly Willowes*, p. 234), had quietly resigned themselves to their lot. In her fictions the woman writer affirmed the importance of workaday lives often thought too dull to merit attention: 'to sit half an hour by an elderly lady getting deaf, another half an hour by some awkward spectacled girl . . . such was generally Mary's fate'.[22] The effect was subtly to shift responsibility, and therefore blame, for her dependency from the spinster herself to her circumstances.

Simply acknowledging the importance of moments of being in the life of a character, who could be represented as wholly pitiful, might in itself be a mark of respect.[23] 'One writes (*one* reason why is) because one does care so passionately that one *must* show it – one must declare one's love',[24] was Katherine Mansfield's tellingly simple reply to John Middleton Murry's praise of her loneliest spinster, Miss Brill. Yet, astonishingly, it is only recently that Mansfield's short stories on the '*dame seule*' theme have been exonerated from the charge of 'cruelty' to their central characters frequently levelled against them in the past.[25] As Kate Fullbrook has argued, recognition of the late short stories as 'unremittingly critical accounts of social injustice grounded in the pretence of a "natural" psychological and biological order' is overdue.[26]

By paying attention to the twists and turns of a spinster's life a literary work need not trade in commonplaces and abstractions about spinsterhood. Nor need it ignore the enriching and worthwhile 'voyages of discovery into new ideas and energetic practices in art or social welfare, education or religion' with which Muriel Spark observed that countless women like the redoubtable Scotish schoolteacher, Jean Brodie, filled 'their war-bereaved spinsterhood'.[27] Sarah Burton, the single-minded teacher in *South Riding*, does not bemoan the lack of opportunities for relationships with men. Their presence would merely distract attention from her work: 'And a good thing too, I was born to be a spinster, and by God, I'm going to spin'.[28] The occupations of schoolteacher or governess provided an attractive vocation for the spinster intent on economic self-sufficiency or on making her influence felt. Indeed, the spinster/companion/governess figure is sometimes the most positive, perceptive, and independent-minded figure in the dramatis personae of Compton-Burnett's fiction. As Marlon B. Ross has observed, to Compton-Burnett, the 'governess represents a way of living, an ideal condition of wise contentedness to which the marginal manless women is best able to aspire'.[29]

Many spinsters' closest and most enduring relationships were with their sisters and the adventures of siblings often provided the staple plots of popular fiction. Deirdre, the narrator of The *Brontës Went to Woolworths*, scoffs at the 'kind of novel . . . called *They Were Seven*, or *Three–Not Out*', in which 'one spends one's entire time trying to sort them all, and muttering, "Was it Isobel who drank, or Gertie?" '[30] Loving and committed relationships between women are dramatized, for example those between Joan Ogden and Elizabeth Rodney in *The Unlit Lamp*, between Muriel Hammond and Delia Vaughan in *The Crowded Street*, and between Averil Kennion and Sally Wraith in *Surplus*. In this forgotten lesbian novel the heroine asserts that 'to limit the fullest manifestation' of the power of love to 'beings, between whom the physical tie of matehood or parenthood exists, is like declaring that electricity can only be generated by one particular kind of dynamo'. *Surplus* ends with Sally wondering whether her role in life is really 'to teach some other unmated woman that she hasn't missed the greatest thing in the world, if she's had a great friendship?'[31]

Many of these women writers rejected the convention of concentrating on one key incident, usually a proposal of marriage, and although marriage is often their families' preferred option it is manifestly not the spinsters'. Indeed for several heroines – Miriam Henderson, Muriel Hammond, Lolly Willowes, Joan Ogden, and Joanna Godden – marriage is rejected precisely because they see it as an imposed rather than a natural condition, a threat to their precious sense of personal autonomy. Flouting convention, Miriam Henderson in *Pilgrimage* plays billiards, smokes in public, wears trousers, and rides a bicycle. She conceives a child with no thought of a wedding ring: 'If you define life for women, as husbands and children, it means that you have no consciousness at all where women are concerned'.[32] For Muriel in *The Crowded Street* to marry Godfrey would be 'to give up every new thing that has made me a person'.[33]

Lolly Willowes feels shy and unnatural in the company of the lawyers whom her in-laws have earmarked as potential husbands. Only well away from the influence of her family is 'the true Laura' able to to discover her real identity, to pour out her soul to her confidant, the devil. As Jane Marcus has noted, *Lolly Willowes* revives the ancient equation of chastity with female freedom and posits chastity not as the absence of experience or sexuality but as a positive state of wholeness for women.[34]

To permit one's heroine to be on such splendidly good terms with the devil, to question and to challenge without hint of punishment, especially in the troubled area of male control and female submission, is to rebel against the dominant social relations. This is irrespective of the extent to which rebellion like Lolly's depended upon an independent income, not just an independent mind, or how circumspect such rebellion may appear today. As Muriel's mentor in *The Crowded Street* tells her, 'the thing that matters is to take your life into your own hands and live it, accepting responsibility for failure or success. The really fatal thing is to let other people make your choices for you'.[35]

The joy of these texts for women who still read them avidly in modern editions is that, once free, there is never a real danger within the narrative that the heroine can ever be boxed back again into the kinds of domestic relationships that define and constrict women's lives. As Lolly Willowes reminds us, 'nothing is impracticable for a single, middle-aged woman with an income of her own' (p. 102). The emphasis of these literary works on the heroine's longing for freedom generates an excess of desire the narrative is at moments hard pressed to contain. In women's fictions, as Nancy K. Miller has argued, the reader's sense of security, itself dependent on the heroine's, 'comes from feeling not that the heroine will triumph in some *conventionally* positive way but that she will transcend the perils of plot with a self-exalting dignity'.[36] Even when a character fails dismally, when the free girl with a burning hope of womanhood becomes, like Joan Ogden, a 'funny old thing with grey hair', contemplating women of the type she had once been, and thinking of herself sadly as 'a forererunner, a kind of pioneer that's got left behind',[37] the novel in question can still be read as a tribute to the vanguard who fell.

The sympathetic literary representation of spinsterhood can perhaps be examined most interestingly in the least explicitly feminist of these women's fictions, *The Rector's Daughter*. This is a highly contradictory text that articulates the problems of the spinster more movingly than the others. But its outlook, consciously outmoded at the time, severely restricts the contemplation of ways of ameliorating her plight. The relationship between Mary Jocelyn and her erudite, but increasingly cantankerous, father is central to this novel. The Rector of Dedmayne is a Victorian patriarch and his unmarried daughter utterly at the mercy of his whims. The village's name is symptomizing: 'dead' plus 'mayne' – a signification between might and main: dead man, deadening power.

But in contrast to May Sinclair, who casts the Vicar of Garth as a paterfamilias with a rod of iron in *The Three Sisters* (1914), Flora Mayor avoids any simple dichotomy between father as oppressor and daughter as victim. It is hinted that Canon Jocelyn's magisterial learning has been paid for heavily by his family. But while Mayor is critical of specific aspects of the Canon's personality, she is also sympathetically aware of the reasons for his intellectual and moral rigidity, namely the intellectual and moral legacy of the world in which he was young, and the querulousness, peccadilloes and infirmity of old age. Jocelyn behaves coldly towards to his handicapped daughter and is indifferent to his absent son, and what Mary feels most deeply of all is her father's lack of feeling. What complicates interpretation of the personal relationship between Mary and her father is their relationship to the wider community in which both have clearly defined public roles.

From the opening pages of *The Rector's Daughter*, Mary Jocelyn is a loved, loving, and altogether indispensable member of her little community: 'In a sense the whole village adored Mary, but quietly' (p. 22). Mary patiently presides over Bible classes, gently officiates at mothers' meetings and harvest festivals, and trains the church choir. Mothers are offended if she is not the first to welcome new arrivals. The Rector's daughter is 'as much a part of her village as its homely hawthorns' (p. 7), adept at patching up the differences between various layers of society which make up the excitement of village life.

Unlike other women novelists of her day, Flora Mayor never questioned the role of family as the basic unit of authority in society. Ivy Compton-Burnett, for example, drew clear parallels between the abuse of power by the head of the household and by the head of the state. 'The assumption of divine right and the acceptance of it takes things further along the same line. History gives us examples that are repeated in smaller kingdoms'.[38] Compton-Burnett is forthright in condemning the power adults exercised over children in the Victorian home: 'I write of power being destructive and parents had absolute power over children in those days'.[39]

But in *The Rector's Daughter* the Jocelyns, father and daughter, serve as a model for the harmonious relationship between the individual and the community in which Mayor strongly believes. This relationship is essentially deferential, supportive of and supported by the allegiances of tradition and hierarchy which the Canon's unstinting devotion to classical scholarship, and Mary's unstinting devotion to the welfare of his parishioners, respectively embody.

Although it is in the home that much of Mary's work is done, it is in the community that her full worth as a human being is recognized. To register the devastating impact of Mary's death from influenza on the tightly knit community to which she belonged the novel changes direction. Narrative progression gives way to retrospective probing to record, subtly and reconditely, the full meaning of this spinster's short life. It was 'crazy' and 'tragic' (p. 336) that Mary did not marry and find happiness in the usual way.

There is an overwhelming sense of sadness and wasted potential in *The Rector's Daughter*, achieved largely through quiet understatement. As a child without friends Mary 'retired within herself, and fell in love instead with Mr Rochester, Hamlet and Dr. Johnson' (p. 14). Largely denied the formal education given to her reluctant older brothers Mary had much time for dreaming and thinking during solitary girlhood walks. 'She had much seething within her waiting for an outlet. She wrote her thoughts down to get rid of them' (p. 19). However, Mary's creativity is never properly recognized. The woman poet is lost – both to herself and to the world – her work dismissed at the end as 'just the Anglican spinster warbling' (p. 337) albeit by a character manifestly incapable of doing it justice.

The cumulative effect of the minutely observed inventory of each every day-to-day task ('solace') that occupies Mary's time in her middle years – winding up clocks, letting the cat and dog in and out of the house – emphasizes the tedium of her existence more tellingly than any protest. No sooner is each duty performed than, as is the punishing nature of such things, it must needs be done all over again. Observation of character ('Aunt Lottie, though not deaf, was inquisitive and inattentive') (p. 325) is precise.

In 'Miss Brill' an impoverished spinster ekes out a living as best she can by giving lessons in English. In Woolf's *Mrs. Dalloway* (1924) Doris Kilman's poverty makes her economically beholden to the employer whom she despises. In *A Room of One's Own* (1928) the narrator survives in the only ways open to a respectable, middle-class woman before 1918: 'addressing envelopes, reading to old ladies, making artificial flowers, teaching the alphabet to small children', 'always to be doing work that one did not wish to do, and to do it like a slave'. One corrosive simile sums up her weary labours: 'like a rust eating away the bloom of the spring, destroying the tree at its heart'.[40]

Mary Jocelyn's unmarried friend, Dora Redland, 'the UF.' or 'Unnecessary Female' (p. 39), paints an idealistic, optimistic gloss on servitude ('the wonderful desire which is so strong in English spinsters to serve, to help' pp. 223–4). But serving others is not enough to make Mary happy. Mary's life is a chastening example of how emptily spinsters have been forced to spend their lives – in the routine exercise of chores that time and duty have drained of meaning, or made the sole repository of meaning. In this respect Mayor shares the concerns of and is writing in conversation with, her sister authors in the 1920s.

The essence of Mary's labour is that it is freely given. At her lowest ebb she forces herself out in the pouring rain to visit a new-born baby: 'But I can't bear the people to think I don't care' (p. 206). Caring of this kind, of the order Mary epitomizes, is clearly intended to make literal and specialized (to women) the qualities of compassion and love: 'Mary had a pull over us in a way', says Brynhilda; 'she cared, and we can't care, not much, and never for long' (p. 336). The far-reaching ethical ramifications of Mary's concern are explained by Mayor's biographer, Sybil Oldfield: 'To refuse to do this work . . . is to force the unwanted failures of the earth to feel the full depth and extent of their unwantedness. It is to kill their spirit'.[41]

A key episode in *The Rector's Daughter* shows how expertly Mayor controls the reader's response. The Rector has delivered a carefully prepared Lenten address to a disappointingly sparse congregation, and has wreaked his irritation on his daughter by contradicting her timid remarks on the way home. In the privacy of her bedroom Mary breaks down in uncontrollable tears – a rare moment at which Mayor allows a real, if momentary, tension between submission to duty and longing for freedom. It is precisely because Mary's patience had hitherto seemed limitless that the surge of emotion between evensong and dinner is so powerful. But even Mary's patience can be stretched to its limits – to breaking point. As the virtuous woman that she is, Mary needs no telling that the married clergyman who is the cause of her anguish is unattainable. But even in cherishing in her heart a forbidden love Mary is allowing herself a liberty that would have shocked her father's generation.

Mayor vividly dramatizes the distress Mary experiences in trying to live in a principled way, feeling deeply and yet, in the main, managing successfully to repress her deepest feelings. For were Mary's emotional needs to be less sensitively drawn, or their containment more mechanical, the degree of sympathetic attention we accord her would be correspondingly less: 'It is like a bitter *Cranford*. Mrs. [*sic*] Mayor explores depths of feeling that Mrs Gaskell's generation perhaps did not know and certainly did not admit to knowing', wrote Sylvia Lynd.[42]

Mary's day-to-day existence amply corroborates Rosaline Masson's view that the home life of the country was based upon 'a substratum of obscure martyrdoms'. But this is not the case that Mayor makes. Personal anguish is not evoked to explore the social structure which makes it feature large in women's lives. On the contrary, it is evoked, as by Charlotte Brontë in *Jane Eyre*, primarily to induce our identification with the heroine in her suffering. Throughout the novel Mary's are the only standards – deeply caring, unselfish, finely discriminating, painfully sensitive, unfailingly generous and compassionate – by which we are asked to judge the behaviour of others.

Cecily Palser Havely, writing on another text, summarizes the general issues which *The Rector's Daughter* also raises for women:

> There can be nothing wrong with any writer's desire to give the fullest possible account of an exclusively female experience or point of view. But the danger in the situation is that so much women's writing about women tends to consolidate the unhappy status quo, and not to advance the claims of women to a fuller life in every respect.
>
> The better the thing is done, the easier it is to believe that this is the thing that women do best, and from there it is too easy a step to believing that it is the only thing they can do.[43]

In analyzing *The Rector's Daughter* I make no criticism of the desire to care for others, quite the opposite. But it is, nevertheless, true that society does not reward 'caring' as it rewards all else it values highly – with money, power, and status – and that society teaches girls to care, and requires this of women, in ways not expected of men. While there is some ironic awareness of this discrepancy in the text ('she knew she excelled in one branch of knowledge – old ladies') (p. 106), the view that caring is women's special province, often used to impede women's progress to full equality, is one the novel in its entirety endorses.

That view is is certainly not that of other novels published in 1924, *The Unlit Lamp*, *Surplus*, *Lolly Willowes*, and *The Crowded Street,* nor is it true of May Sinclair's earlier *Life and Death of Harriett Frean.* The heroine of *Surplus*, Sally Wraith, who sets up a motor vehicle business, admits to having as much inclination and aptitude for work with the sick and the poor 'as the pirate chief has for assuming the cassock'.[44] In the beech forests of the Chilterns, Lolly Willowes is determined to resist the temptation to do good which pursues her relentlessly. She emphasizes that she did not opt to become the village witch in order 'to run round being helpful', to become 'a district visitor on a broomstick' (pp. 238–9), but to 'escape all that – to have a life of one's own, not an existence doled out to you by others' (p. 239).

Radclyffe Hall and Winifred Holtby do not consider it sufficient to speak up for women. They plumb the source of women's problems in the expectations of the family. Muriel Hammond in *The Crowded Street* – originally called *The Wallflower* – turns her back on the tedium of her life at home in the suburb of Marshington, in Cottingham, where Holtby herself grew up, to work for a women's reforming organization in London, to put into practice 'an idea of service – not just vague and sentimental, but translated into quite practical things'.[45]

In *The Unlit Lamp*, the personal attention Joan Ogden's selfish, ailing mother demands effectively wrecks her daughter's promising career as a doctor. The irony of medicine never being recognized as caring work (because it falls outside the ambit of traditional work for women) is heavily underscored. Try as she might Joan cannot free herself from the kindly tyranny of family interest and in the end is forced to give up not just the hope of a career but also the hope of establishing any kind of independence. Like Lalage Rush in *Ordinary Families* (1933) Joan can offer no convincing reason for wanting to leave home. In *Ordinary Families* the narrator tartly notes how 'in very few of the thousands of good homes, from which the children struggle to escape, can the truth ever be told ... even in the heat of a family row'.[46]

Lolly Willowes, The Unlit Lamp, and *The Crowded Street* are a good deal more critical of the middle-class family's expectations, and a good deal more scathing about domestic obligations imposed upon the unmarried daughter, than is *The Rector's Daughter*, in which there is a strong suggestion that Mary would have made a good wife and a loving mother, but little suggestion – if we discount her secret writing – that she might have become anything else.

One question that arises is why Mayor conflates the important work of caring for others – which most readers would agree is genuinely life-enhancing – and the routine performance of household duties which may appear to the reader as merely domestic drudgery passed off as caring in a thin disguise. If, as would appear to be the case, we are expected to respect the selfless way that Mary performs *all* the work required of her, why is this dedication to domestic duty so frequently lauded in literature only when exemplified by women? The truth is that *The Rector's Daughter* is profoundly ambivalent on the question of self-sacrifice. As Merryn Williams succinctly puts it, Mayor did not want to 'denounce the traditional Christian and womanly virtue of self-sacrifice, although she did not recommend it to all and sundry'.[47]

Mayor admits that the psychological toll extracted from Mary as a 'dutiful daughter' is too high, but she will not generalize from that to the patriarchal system. In the end, the paternal order that the novel embodies both establishes an authoritative system of values and abolishes the apparent discrepancy between individual desire and social responsibility. What is absent in *The Rector's Daughter*, but markedly present in other texts about spinsters of this period, such as *Lolly Willowes*, *Life and Death of Harriett Frean*, and *The Crowded Street*, is the potential for changing the ground rules of domestic attitudes and for questioning the relationships between women and men. Mayor is out of step, and probably out of sympathy, with other women writers who want to question these values or who may suggest that some restructuring of society's expectations of women may be desirable.

In the matter of the Lenten sermon we have seen the clergyman depicted at his worst, but the portrait of Canon Jocelyn is not unmitigatedly severe, and dexterous touches soften the general picture of unbending rectitude. Despite his lack of feeling for his other children, he is genuinely fond of Mary, and tries to console her when he senses her disappointment in love. But he cannot help patronizing even when well-intentioned: ' "It's an excellent wine, and" (in the warmth of his feeling he said "and" not "but") "it is particularly well suited to ladies" ' (p. 143). A journal entry of her father's expressing remorse over the death of his younger daughter, Ruth, which Mary discovers after his death – 'I feel no grief. How should I? I have not deserved to grieve' (p. 313) – presents the Canon in a more sympathetic light.

Selective details of this kind generate the novel's representation of values. They humanize the Rector and rub out any lingering suspicion that Jocelyn is really little better than 'an ogre father battening on his daughter's vitality' (p. 217), subtly erasing any trace of substantive conflict that may once have existed between Mary and her father. Whatever their differences, we are left in no doubt that they are cast in the same mould, and that *both* epitomize all that is worth cherishing in the little community: 'The Jocelyns must have been wonderful, and the people were dreadfully upset when they heard she had passed away' (p. 335).

There is an elegaic, contemplative tone to *The Rector's Daughter*, and the first impressions that we have of Dedmayne in this extended rural retrospect are the ones that remain with us. The emphasis is on a vanishing world: 'what has been known from childhood must be lovable, whether it is ugly or beautiful' (p. 6), and on an old-fashioned little community, just on the verge of extinction, its fragility sensitively evoked. Mary's home, the Rectory, is a 'frail, frail survival, lasting on out of its time, its companions vanished long since' which would 'fall at a touch' (p. 12) when the Canon himself died. The middle-aged feel the Canon to be 'a belated traveller' (p. 317). Mrs Plumtree is a 'faded specimen of the generation that is almost gone' (p. 106).

The parishioners, Miss Gage and Mrs Davy, are finely drawn, their tiny ambitions and harmless affectations gently ridiculed, but only from a perspective that is essentially affectionate. Any humour derived at their expense is, like Elizabeth Gaskell's in *Cranford*, always an expression of the author's enjoyment of their idiosyncrasy and of her desire to leave the objects of her humour unchanged.

In throwing a ladder back to pre-1914, to an isolated hamlet where time stands still, Mayor denies Mary options she knew were open to educated women in the 1920s. Her avoidance of the difficult questions that Freudian theory and feminism, in their different ways, were raising for women marks Mayor's difference in outlook from other women writers. The expectations and life of Mary Jocelyn correspond to the social position and subjective experiences of women in an earlier age. This is essential to the realization of a caring society and the avoidance of vexed issues that might otherwise have to be confronted. For the relationship of a text to the values and ideas of its time is illustrated by what it says and by what it does not.

It is in the 'significant *silences* of a text, in its gaps and absences, that the presence of ideology can be most positively felt'.[48] The real effect of dwelling upon a caring society in the past, as Raymond Williams astutely points out in *The Country and the City*, is often withdrawal from a full response to existing society. 'Value', he writes, 'is in the past, as a general retrospective condition, and is in the present only as a particular and private sensibility, the individual moral action'.[49]

At one level of argument a text illustrates nothing beyond itself. To value it for its referential qualities, as a response to social realities, is to say that it has failed to transcend its material. 'Real art', as Susan Sontag explains, always 'has the capacity to make us nervous. By reducing the work of art to its content and then interpreting that, one tames the work of art'.[50]

But to understand the relationship of any work of art to the ideologies current at its moment of production is a different objective. This must involve closely examining the specific material conditions, the actual relations, and the deeper contradictions out of which both works of art and specific ideologies have emerged. *The Rector's Daughter* emerged in the context of the national preoccupation with two million supposedly superfluous women. But here lie problems for the feminist reader. For many of the impulses informing the text, particularly in relation to its defence of family, the position of women, and the kind of society Mayor upholds, are deeply conservative.

The major achievement of *The Rector's Daughter* is to create a heroine who, on the surface, appears nothing more than a garden-variety Anglican spinster, but whose life is acknowledged to be exemplary by the community in which she lives. In the context of the dominant public attitudes to the spinster in the 1920s this is significant. Mary is clearly in no way 'superfluous' or 'useless' or 'abnormal'. The radical potential of this representation is, however, contained and attenuated by Mayor's affirmation of the importance of traditional family values and of heterosexual, romantic love.

The ideological project of *The Rector's Daughter* is to stress the need for personal responsibility in a changing world. To do this Mayor must either simplify or exclude that world. For, as Raymond Williams observes, it is a contradiction in the form of the novel (as received and developed by George

Eliot) that the 'moral emphases on conduct – and therefore the technical strategy of unified narrative and analytic tones – must be at odds with any society – the "knowable community" of the novel – in which moral bearings have been extended to substantial and conflicting social relationships'.[51]

The only values offered that jar with those of Dedmayne are voiced by Brynhilda Kenrick and her fashionable circle who had 'shaken off their families and united in light elastic unions with friends' (p. 95). Wives without husbands and husbands without wives – this travesty of the Bloomsbury circle could easily be recognized and dismissed by the first readers of the work. It is the only challenge to the caring community admitted to the work – and no real challenge at all.

The Rector's Daughter is a product of the anxieties besetting Mayor's class, and the social milieu represented in the novel is similar to her own. It may be usefully contrasted with D.H. Lawrence's 'Daughters of the Vicar' (1914) in which Lawrence describes a vicar's family isolated and surrounded by working-class people who have scant regard for their social superiors. The church at Aldecross stands '*buried* in its greenery, *stranded* and *sleeping* among the fields, while the brick houses elbow nearer and nearer, *threatening* to crush it down' (my emphases).[52] This short story indicates Lawrence's own keen awareness of the existence of communities which may be unknowable to outsiders and deeply threatening to middle-class sensibilities. We have only to put it beside *The Rector's Daughter* to be made aware of that novel's silences in relation to class and to sexual politics, and of the social tensions that Flora Mayor refused to face.

Today we see the category of 'the frustrated spinster' as a fabrication, a severely limited, deficient work of definition, but in the 1920s it passed for truth. At a time when the popular perception of the spinster was deceptively simple: 'all those girls are just as like one another as two peas', a matron observes with 'married hardness' in *The Rector's Daughter* (p. 30), the very diversity of literary representation in the 1920s introduced a welcome complexity. There is much in these novels to suggest that their heroines are typical. May Sinclair, for example, in framing her heroines' predicaments within the paradigms of Freudian theory, makes it clear that what is true for them is true of other women of their backgrounds. But there is also much that focuses on what is atypical and unique.

Women writers felt personally the weight of ideologies which constantly threatened to reduce the spinster to the status of object. Their response, in the idiom of the nineteenth century, took the form of a 'covert solidarity that sometimes amounted to a genteel conspiracy' between the writer and her reader. In the idiom of the present we may choose to refer to this as sisterhood. Specifically, what Elaine Showalter has termed as 'active unities of consciousness'[53] are manifested in the empathy between the woman writer who had broken free from convention and her character whose life is still hemmed by it: 'There was a moment when I first had "the idea" when I saw the two sisters as *amusing*; but the moment I looked deeper (let me be quite frank) I bowed down to the beauty that was hidden in their lives and to discover that was all my desire',[54] wrote Katherine Mansfield of Josephine and Constantia Pinner. Not all the authors to whom I have referred were spinsters. But with the exception of Mansfield and Richardson, those of most interest today – Compton-Burnett, Hall, Holtby, Mayor, Sinclair, and Townsend Warner – wrote from first-hand experience. Moreover, the desire of these women to speak from their own personal knowledge of spinsterhood occasionally slips into their texts in untoward ways. Mary Jocelyn is not Flora Mayor's surrogate, despite the fact that her own father was a clergyman of some intellectual distinction. But one clergyman's daughter's understanding of the heartache of another (the 'odd cry from the heart, or whatever there is beyond the heart' – p. 337), permeates the novel. At times each of these authors appears to be working through some painful personal response to external circumstances, to be tracing or contemplating her own flight path of escape: 'My Muriel is myself – part of me only – the stupid frightened part', wrote Winifred Holtby.[55]

Individually, as literary works, only *The Rector's Daughter*, *Lolly Willowes*, Dorothy Richardson's *Pilgrimage*, 'Miss Brill', and 'The Daughters of the Late Colonel' are of significant interest. But read in context, the women's novels of the 1920s acquire new meanings as symptoms of women's unwillingness to submit to those values, images, and ideas, which ensure either that the dominant social relations are seen by most people as 'natural' or are not seen at all. And collectively, when each work is read with reference to the others, they delineate a form of resistance, albeit at times hesitant and ambivalent, to the distorting ideologies of the day.

Notes

1   K. Mansfield (1945), *The Collected Short Stories*, London, Constable, pp. 335, 336.
2   J. Lewis (1984), *Women in England 1870–1950: Sexual Divisions and Social Change*, Brighton, Wheatsheaf, p. 4.
3   B. Melman (1988), *Women and the Popular Imagination in the Twenties: Flappers and Nymphs*, Basingstoke, Macmillan, p. 18.
4   'Woman', *New Statesman*, 24 September 1921, p. 669.
5   Quoted in H.D. Zegger (1976), *May Sinclair*, Boston, Twayne, p. 128.
6   V. Brittain (1933), *Testament of Youth*, London, Gollancz, pp. 577, 578.
7   U. Troubridge (1961), *The Life and Death of Radclyffe Hall*, London, Hammond, p. 69.
8   L. Gordon (1977), *Woman's Body, Woman's Right: A Social History of Birth Control in America*, Harmondsworth, Penguin, p. 194.
9   Sheila Jeffreys (1985), *The Spinster and Her Enemies: Feminism and Sexuality 1880–1930*, London, Pandora, p. 155.
10 See M. Hunt (1990), 'The De-Eroticism of Women's Liberation: Social Purity Movements and the Revolutionary Feminism of Sheila Jeffreys', *Feminist Review*, 34 (Spring), pp. 18–23.
11 W. Holtby (1932), *Virginia Woolf: A Critical Memoir*, London, Wishart, p. 29.
12 Letter to Beatrice Campbell, dated May 1916, in C.K. Stead (Ed.), *The Letters and Journals of Katherine Mansfield: A Selection*, London, Allen Lane, 1977, p. 76.
13 Holtby (1932), p. 30.
14 A.M. Ludovici (1925), *Lysistrata; or Woman's Future and Future Woman*, London, Kegan Paul, Trench, Trubner, p. 37.
15 A.M. Ludovici (1936), *The Future of Woman*, London, Kegan Paul, Trench, Trubner, p. 70.

16   A.M. Ludovici (1923), *Woman: A Vindication*, London, Constable, p. 231.
17   D. Sayers (1935), *Gaudy Night*, London, Gollancz, p. 81.
18   'Dark Stars (Unpaid), 2, Unmarried', *Time and Tide*, 11 March 1921, p. 227.
19   Sylvia Townsend Warner (1926), *Lolly Willowes; or the Loving Huntsman*, London, Chatto and Windus, p. 2. All quotations are from the first edition and page references are given in the main body of the text.
20   M. Sinclair (1922), *Life and Death of Harriett Frean*, London, Collins, p. 99.
21   *Ibid.*, p. 110.
22   F.M. Mayor (1924), *The Rector's Daughter*, London, Hogarth, p. 82. All quotations are from the first edition and page references are given in the main body of the text.
23   R. Williams, ' "Hit or Miss": The Middle Class Spinster in Women's Novels 1913–1936', unpublished part 2 tripos dissertation, Cambridge University, 1985.
24   Letter to John Middleton Murry, dated 21 November 1920, in J.M. Murry (Ed.), *Katherine Mansfield's Letters to John Middleton Murry: 1913–1922*, London, Constable, 1951, p. 598.
25   See, for example, M. Drabble (1973), 'Katherine Mansfield: Fifty Years On', *Harpers and Queen*, July, pp. 106–7.
26   K. Fullbrook (1986), *Katherine Mansfield*, Brighton, Harvester, p. 127.
27   M. Spark (1961), *The Prime of Miss Jean Brodie*, London, Macmillan, p. 52.
28   W. Holtby (1936), *South Riding: An English Landscape*, Collins, p. 57.
29   Marlin B. Ross (1991), 'Contented Spinsters: Governessing and the Limits of Discursive Desire in the Fiction of I. Compton-Burnett', in Laura A. Doan (Ed.), *Old Maids to Radical Spinsters: Unmarried Women in the Twentieth-Century Novel*, Urbana and Chicago, University of Illinois Press, pp. 39–65, at p. 63.
30   R. Ferguson (1931), *The Brontës went to Woolworths*, London, Ernest Benn, 1931, p. 7.
31   S. Stephenson (1924), *Surplus*, London, Fisher Unwin, pp. 294, 313.
32   D. Richardson (1921), *Pilgrimage* (13 vols), vol. 3, *Deadlock*, London, Duckworth, p. 298.
33   W. Holtby (1924), *The Crowded Street*, London, John Lane, p. 313.
34   J. Marcus (1984), 'A Wilderness of One's Own: Feminist Fantasy Novels in the Twenties: Rebecca West and Sylvia Townsend Warner', in S.M. Squier (Ed.), *Women Writers and the City: Essays in Feminist Literary Criticism*, Knoxville, University of Tennessee Press, p. 135.
35   W. Holtby, *The Crowded Street*, p. 261.
36   Nancy K. Miller (1981), 'Emphasis Added: Plots and Plausibilities in Women's Fiction', *PMLA*, vol. 96, no. 1 (January), p. 40.
37   R. Hall (1924), *The Unlit Lamp*, London, Cassell, p. 301.
38   'I. Compton-Burnett and M. Jourdain – a Conversation' (1934), *Orion*, 1, reprinted in C. Barton (Ed.) (1972), *The Art of I. Compton-Burnett: A Collection of Critical Essays*, Gollancz, pp. 21–31, at p. 29.
39   Quoted in H. Spurling (1984), *Secrets of a Woman's Heart: The Later Life of Ivy Compton-Burnett 1920–1969*, London, Hodder and Stoughton, p. 174.
40   V. Woolf (1929), *A Room of One's Own*, London, Hogarth, p. 57.
41   S. Oldfield (1984), *Spinsters of this Parish: The Life and Times of F.M. Mayor and Mary Sheepshanks*, London, Virago, p. 237.
42   *Time and Tide*, 18 July 1924, p. 691.
43   'Carson McCullers and Flannery O'Connor' (1982), in D. Jefferson and G. Martin (Eds), *The Uses of Fiction: Essays on the Modern Novel in Honour of Arnold Kettle*, Milton Keynes, Open University Press, p. 118.
44   S. Stephenson, *Surplus*, p. 309.
45   Holtby (1924), p. 305.
46   E. Arnot Robertson (1933), *Ordinary Families*, London, Jonathan Cape, p. 217.

47 M. Williams (1987), *Six Women Novelists*, Basingstoke, Macmillan, p. 52.
48 T. Eagleton (1976), *Marxism and Literary Criticism*, London, Methuen, p. 35.
49 R. Williams (1973), *The Country and the City*, London, Chatto and Windus, p. 180.
50 S. Sontag (1969), *Against Interpretation and Other Essays*, London, Eyre and Spottiswoode, p. 8.
51 R. Williams, *The Country and the City*, pp. 168–9.
52 'Daughters of the Vicar', in *The Tales of D.H. Lawrence* (1934), London, Martin and Secker, p. 47.
53 E. Showalter S. (1987), *A Literature of their Own: British Women Novelists from Brontë to Lessing*, London, Virago, pp. 15–16.
54 Letter to W. Gerhardie, 23 June 1921, in J.M. Murry (Ed.) (1928), *The Letters of Katherine Mansfield*, 2 vols, London, Constable, vol. 2, p. 120.
55 Letter from Holtby to 'Rosalind', dated 28 October 1924, in A. Holtby and J. McWilliam (Eds) (1937), *Letters to a Friend*, Collins, p. 288.

Chapter 17

Chloe, Olivia, Isabel, Letitia, Harriette, Honor, and Many More: Women in Medicine and Biomedical Science, 1914–1945

Lesley A. Hall

'Chloe liked Olivia. They shared a laboratory together', wrote Virginia Woolf in *A Room of One's Own*, evoking the imaginary contemporary novel *Life's Adventure*, in which 'these two young women were engaged in mincing liver, which is, it seems, a cure for pernicious anaemia; although one of them was married and had – I think I am right in stating – two small children'.[1] Jane Marcus has flatly declared that 'Woolf's dream of a married woman working in a lab is Utopian',[2] but her assumptions, not Woolf's, are at fault. Woolf's essay was originally presented to an audience at Girton, which had, as she surely knew, produced many distinguished women scientists.[3] It was not wholly improbable that a married woman, even a mother, might be mincing liver in a laboratory. Woolf was personally acquainted with Janet Vaughan, who developed her own extract of raw liver for treating pernicious anaemia (and continued to work throughout her daughters' childhood),[4] while Dorothy Hodgkin, a married woman with children as well as a Nobel-Prize-winning crystallographer, elucidated the crystalline structure of Vitamin B12: its subsequent synthesis meant that victims of pernicious anaemia no longer had to choke down quantities of raw liver.[5]

This chapter does not pretend to present a definitive narrative of its topic, or even to reclaim huge numbers of women doctors and scientists formerly 'hidden from history', but argues that there are a great many stories about women in medicine and the biomedical sciences which have not yet been told. It will consider the patterns shaping women's careers in medicine and the biomedical sciences during this period. We know much about the heroic age of the first professional women doctors, and something of those medical women who went to the Balkans and France during the Great War. But few of the many interesting, if less immediately dramatic, stories of the inter-war years have yet been told, which do not merely recover hidden heroines but suggest further lines of inquiry into the possible relations between women and science.

These relations are undoubtedly problematic, given the construction of 'science' as a masculine endeavour: though recent feminist historians of science have pointed out the variations between different disciplines in their capacity to 'accommodate feminization'.[6] This essay looks at women whose engagement with their chosen fields illuminates female narratives of the medical and scientific career: neither mere feminine versions of triumphal male tales of the Wonderful Onward March of Medical Progress (WOMOMP), nor exemplars of victimization of women by the masculine power of medicine and science.

Women in Medicine

There appear to be significant generational differences between women qualifying before the World War One and after it, a dichotomy which can possibly be pushed too far. The heroic age was past; women's right to medical education and employment as doctors had been established. Though fifteen hospitals were founded by and for women in the United Kingdom and India between 1866 and 1916, only one more, the Marie Curie in North London (1929, for radiotherapy in uterine cancer), was founded after that date.[7] Possibly wider opportunities for women in the medical field obviated the need for special institutions: however, special institutions may have been taking a different, less tangible form than the bricks and mortar of hospitals, in a world of change. The following case histories are not in any way 'typical', but illustrate some of the diverse paths a woman's career in medicine could take.

By the 1920s women over 30 had obtained parliamentary suffrage, and certain sex disqualifications had been removed. However, Dr Isabel Hutton found, following service in the Balkans and Turkey during and just after the War, that her marriage to an Army officer constituted 'a bar to salaried public health and hospital appointments. Women could, of course, take up general practice, but many wished to continue what they had been specially trained for'.[8]

In many ways Dr, later Lady, Hutton (*c*. 1888–1960) enjoyed certain advantages. Her supportive husband accepted a semi-detached marriage while she built up a practice, their marriage was childless, and she had acquired considerable career experience in psychiatric medicine prior to her marriage, including study in Munich and Vienna. Sir Frederick Mott, one of her lecturers on a postgraduate course on shell-shock at the Maudsley Hospital, offered Hutton a research post (funded for a year by the Board of Control) in his 'familial' laboratory. However, hearing she was married, the London County Council refused to admit her to the established staff, not even examining her credentials. The Government post of Junior Commissioner in Lunacy was similarly closed to her on account of her marital status. To the Secretary of the Board of Control she 'replied in an angry staccato that it was a pity I had disclosed this heinous crime of marriage ... better ... would it have been to live in sin and then all posts would have been open to me'.[9] She therefore set up in consulting practice, while continuing laboratory research, and finally attained an honorary consultant's post (for which 'few men would think of applying') at the 'obscure and shabby' British Hospital for Mental and Nervous

Disorders. This institution was 'unorthodox, almost unknown, and had no prestige': but she was overjoyed to 'have my own clinic and carry out treatment on my own lines', and stayed for thirty years.[10]

Amongst other activities Hutton was on the subcommittee of the Medical Women's Federation investigating women's ineligibility for commercial pilots' licences on the grounds that menstruation and pregnancy rendered them 'medically unfit', and was also MWF representative on the British Medical Association's Psychoanalysis Committee. When her husband's army career took them to India, Dr Hutton ended up as Director of the Indian Red Cross Welfare Service during World War Two.[11] This versatility in turning her hand to whatever happened to offer in the situation in which she found herself, as opposed to planning out and following a definite career path, is perhaps a peculiarly female trajectory. However, in spite of the professional satisfaction Hutton undoubtedly achieved in the face of unpromising circumstances, her case illuminates the difficulties the most dedicated and determined woman faced in wishing to combine marriage and a medical career, even in one of the less glamorous specialities.

Remaining unmarried had distinct advantages for a woman who wanted a career with administrative and policy-making responsibility. Dr Letitia Fairfield (1885–1978) served most of her life with the London County Council which had rejected Isabel Hutton. Qualified in both law and medicine, a suffragette and a life-long member of the Fabian Society, Fairfield initially seems almost one of H.G. Wells' 'Samurai' come to life. However, in other respects she was far from a Wellsian woman, regarding his liaison with her younger sister Rebecca West as unedifying in the extreme, and herself refusing both marriage and motherhood and converting to Roman Catholicism. Her career followed a path almost comparable to that of a male doctor of her abilities and qualifications, even to the form of interruption caused by the two world wars.

Fairfield wielded considerable public influence; joining the London County Council in 1911, she eventually became its first woman senior medical officer. During the Great War she was medical area controller attached to the Women's Army Auxiliary Corps and later inspector of medical services, Women's Royal Air Force, and again active in the military medical services for women in World War Two. Her civilian interests, medically speaking, seem to have been in what might be assumed to be 'feminine' areas – mother and child welfare, child guidance, care of the elderly – as well as venereology: in spite of its enduring stigma, an exciting field at this period. She was a founder-member of the Medico-Legal Society, and was active in the Medical Women's Federation, becoming its president, 1930–1932.

Fairfield was no bland bureaucrat: 'a delightful, vigorous, dogmatic and sometimes infuriating companion' who did not 'always see eye to eye with her colleagues', she was nevertheless a 'much appreciated medical administrator' whose 'integrity and loyalty were everywhere appreciated'. Neither was she hobbled by petty fears of inconsistency; after many years of publicly opposing all forms of birth control (in tune with her Roman Catholic faith) in later life Fairfield supported campaigns to change the church's teaching.[12]

A woman of a generation junior to these pioneers was Joan Malleson (1900–1956). Educated at the pioneering coeducational establishment Bedales, even before qualifying in medicine she married the actor Miles Malleson – possibly a new, post-war, 'having-it-all' concept of the professional woman's life – though they eventually divorced (their two sons also became doctors). She is principally notable for belonging to that 'select and courageous group of women doctors who were pioneers in sexual reform' whose 'views and practice, considered heretical and shocking at the time', by the 1950s became 'widely accepted as conventional and orthodox'. She was involved in setting up the Family Planning Association, and was instrumental in disseminating both the desirability of birth control and reliable methods. One of the first doctors to run a clinic for sexual difficulties in conjunction with family planning work, she also trained medical students at University College Hospital in contraceptive techniques, and wrote a number of works of advice for a lay audience under the pseudonym 'Medica'. Although it is known by the name of Alec Bourne, the gynaecologist who performed the operation, Malleson was the moving force behind the famous 1938 case creating common-law precedent enabling doctors to perform abortions for psychiatric reasons.[13]

The Family Planning movement provided opportunities for many women doctors (and for nurses and lay-workers) who for family reasons were unable to pursue a full-time career. They were able to do fixed sessions at agreed times, and while there may not have been much career progression, they were in touch with the fields of gynaecology, obstetrics, and psychiatry as well as with developments in birth control and marriage guidance.[14]

There were obviously a variety of career patterns possible to women doctors at this period.[15] One area very hard to uncover would be the role of women in general practice: for complex reasons including the legal status of their records the documentation of general practice generally is not easy to come by. However, records of hospitals and of organizations in the medical and social welfare area, as well as records of local authorities relating to mother and child welfare work, and those dealing with Poor Law and municipal medical institutions at which unglamorous end of medical practice women were often employed, would repay investigation.

Women seldom obtained positions at the most prestigious teaching hospitals. However, good work could be done at institutions which were not the glamorous flagships of medical science. Mary Walker (1888–1974) spent her entire career in 'the academic wilderness' of Poor Law (later municipal and National Health Service) hospitals in South London. Although such institutions did not 'stimulate or nourish the questing spirit', she was responsible for 'The Miracle at St Alfege's', discovering the value of physostigmine for treating the debilitating muscular disorder myasthenia gravis. Her theory, first propounded in a letter to *The Lancet* in 1934 and a year later in a paper to the Royal Society of Medicine, as her doctoral thesis won her the Edinburgh Gold Medal in 1937. 'Relegated to ... a lowly post at St Francis Hospital', she was denied opportunities to continue research, since patients from outside the catchment area, or ineligible for Poor Law treatment, could not be referred to her. Dependence upon her full-time salaried job meant that she was unable to accept an honorary post at the Elizabeth Garrett Anderson Hospital with patient beds, though this shy and retiring 'great medical discoverer' could have established a flourishing private practice in neurology.[16] Was it timid lack of confidence that kept Dr Walker in her humble salaried post, or were there other factors, for example family responsibilities, which deterred her from the risk?[17]

Women in Science

Patricia Phillips has painted a depressing picture of actual decline in the possibility of satisfying involvement for women in science, as amateur household or craft tradition shifted to institutionalization and professionalization.[18] However, her vision of women scientists in the early twentieth century as second-class intellects deemed unfitted to study the classics, doomed to a life of schoolteaching on the model of the pathetic Aggie Sigglesthwaite in Winifred Holtby's *South Riding*, is rather too gloomy, and rather underestimates the ways women can 'twist male-imposed ideologies of science to their own ends'.[19] The Household and Social Science Department at King's College for Women, 'founded to provide an education in household science with a genuinely scientific basis and a full academic status', was seriously committed to scientific rigour, appointing Edward Mellanby (later Secretary of the Medical Research Council) Professor of Physiology in 1913. The Dean, Janet Lane-Claypon, was a physiologist in her own right who collaborated with eminent (male) researchers and did significant work on the physiology of lactation.[20] The ideology of 'domestic science' may have reassured men, and anxious parents, while permitting women genuine engagement with laboratory science.

In order to retrieve the stories of women in science we must turn away, as Marina Benjamin has pointed out, from the focus on 'elite, high-class' institutionalized science and the 'hard' sciences.[21] And as Pnina Abir-Am and Dorinda Outram have suggested, the male-orientated model of 'long uninterrupted hours of concentrated work in the field or the laboratory' may not fit all scientists; 'through ingenuity, careful planning and a willingness to defy social convention' some female scientists combined 'intensive work in science with time for family duties and pleasures'.[22] The criteria of 'success' ought also to be queried: women may publish less than their male colleagues but is quantity necessarily related in a simple ratio to quality?

Some scientific disciplines were notable for including women from an early stage, for example biochemistry and crystallography, which share certain characteristics. They were borderline disciplines, developing fields with 'entry paths and entry qualifications ... not well-defined',[23] possibly unattractive to men able to make a career in established fields, whereas even 'first-class women' had less generous opportunities. These disciplines were characterized generally by an 'unaggressive low-key friendly atmosphere' emphasizing cooperation over competition.[24] Male mentors prepared to encourage and support women as colleagues were an important factor – Gowland Hopkins for the early Cambridge biochemists, the Braggs, father and son, and J.D. Bernal with the crystallographers: men themselves to a certain degree off the mainstream of the stereotyped establishment career.[25]

Within fields still in the process of establishing themselves there were few, if any, salaried academic posts available, and early biochemists of both sexes were often obliged to support themselves doing applied research in hospitals, sanitary commission laboratories, and research institutes, environments in which at least they might contrive to pursue their own research. Harriette Chick, for example, had been an assistant bacteriologist to the Royal Commission on Sewage Disposal before receiving the Jenner Memorial Research Studentship at the Lister Institute.[26]

The funding of posts for women in particular, however, was uncertain and often year by year through awards from various grant-making bodies, although Muriel Robertson said of Marjory Stephenson, one of the first two women to receive the Fellowship of the Royal Society,[27] that it could have been 'a sign of the distrust of a somewhat new subject that so original a worker should have been an annual grantee for so many years' rather than a strictly gender issue. Dorothy Needham 'simply existed on one research grant after another, devoid of position, rank or assured emolument'. Women in biochemistry were characteristically independent research workers 'not competing for salaried positions despite life-long commitment to first class scientific work', receiving professional recognition from their peers but seldom considered for established posts.[28]

The place of funding bodies, both private philanthropic organizations such as the Beit Memorial Fellowship Trust, and more official bodies such as the Medical Research Council, in women's careers in science has not been fully explored. The role of Sir Edward Mellanby in promoting women in medical research, especially during his years as Secretary of the Medical Research Council, from 1933 to 1949, has not yet been studied. His early connection with King's College for Women has been mentioned and his post-retirement collaboration with Honor Fell is discussed below. His wife, May Mellanby (née Tweedy), was both a collaborator enabling him to combine research work with an administrative career, and a noted researcher into the physiology of dentition in her own right. It is, of course, possible that in emergent fields like biochemistry there was no large pool of talent to draw on, and therefore conceivable that awards were likely to go to whoever was capable of doing the work: married women, even mothers.

From very early on – the fifth fellowship awarded went to Ida Smedley – women received Beit Memorial Fellowships for research work, and, considering the proportion of women to men in science, a considerable percentage. Out of 236 fellowships awarded up to 1936, thirty-five went to women – nearly 15 per cent – among recipients including some of the biggest names in the WOMOMP. It is possible that it was unrepresentative of funding bodies generally, fellowships being also granted to men who did not pursue linear research careers but took time out from general practice or public health work. A few interesting general points about these thirty-five female Beit Memorial Fellows: at least one had obtained her first BSc in Household and Social Science from King's College for Women. One was a retired Science Mistress from Croydon Girls' School, another was the principal of a Horticultural College. One was the Director of Research in Bee Disease to the Ministry of Agriculture. Six had some connection with the Lister Institute. Two had been engaged in psychotherapeutic practice. Dorothy Needham had been on the British Scientific Mission to China, in spite of her aforementioned lack of any established position. At least half were married at some stage of their career.[29]

Access to grants depended to a considerable extent upon networks of patronage: attitudes to women of learned societies played a role not yet fully investigated. The Physiological Society did not officially exclude women, nevertheless no women became members between 1876, when it was founded, and 1915, although women gave presentations at meetings during that time. Their admission was the subject of a two-year debate within the Society. The Biochemical Society, however, though initially specifically excluding women, rescinded this provision very early on.[30]

The role of private research institutions such as the Lister Institute, less constrained by the traditions, formal and informal, of longer established and more secure bodies, also needs further exploration. The first woman to be appointed at the Lister (in spite of opposition in some quarters) was Harriette (later Dame Harriette) Chick in 1905. Trained as a bacteriologist, Chick moved into biochemistry and in particular the biochemistry of nutrition, a shift from field to field instead of linear progression as typical of women's scientific careers as it was in medicine. Following Chick's appointment a number of very competent women were appointed to the Lister or spent time working there.[31]

The Lister was at least on a sound financial footing following an endowment from Lord Iveagh in 1898. The Strangeways Research Laboratory in Cambridge, founded in 1905 by T.S.P. Strangeways as the Cambridge Research Hospital, ran from year to year on a shoe-string, sustained by grant-funding. Following Strangeways' unexpected death in 1926, the Trustees of the Laboratory appointed as Director a young woman who had been working with him (supported by a Beit Fellowship) on the development of tissue culture, Honor Bridget Fell (1900–1986), later to become a Fellow of the Royal Society and a Dame of the British Empire. She remained Director until 1970, when she retired from administrative duties but continued her research.

Fell is a fascinating figure. For forty years she ran the laboratory with an absolute minimum of support staff, without salary – she was paid by a Royal Society research grant. (While women were not admitted to the Fellowship of the Society before 1945, it was financially supporting women scientists from at least the turn of the century: this needs further investigation.) Fell thus had unusual importance as an organizer and promoter of scientific work by others. Under her Directorship the Strangeways became renowned worldwide, scientists as far away as the USA and Japan begging to work there and not infrequently having to be put off because of limited space. Descriptions of the 'laboratory culture' of the Strangeways convey an atmosphere almost familial – with the entire staff coming together at tea-time to discuss their work informally. It was remarked similarly of the laboratory of Dorothy Hodgkin that 'despite being a gloomy place' it was nonetheless 'happy and homely'.[32] This suggests that women heading scientific teams did not require an aggressively male and hierarchical style of leadership to run fruitful laboratories, although this would seem partly to relate to the styles in particular disciplines, being characteristic also of laboratories under Gowland Hopkins and Bernal.

Running a laboratory might be considered a more than full-time job; but Honor Fell was also a distinguished scientist who did much important research in and with tissue culture. While – one of the advantages of remaining unmarried! – she was able to pursue research at weekends, her success in integrating running an institution with a productive research career owed much to 'the extraordinarily integrated quality of her work pattern'. A colleague described visiting the lab one Sunday, and being invited by Fell to join her for morning tea, prepared by herself: 'prior to tea, culture dishes had been set out and media prepared. After tea the embryonic tissues were to be dissected ... the slight increment of humidity that resulted as a dividend of the tea ceremony was enough to minimize the desiccation problem' normally complicating this task; a small example of Fell achieving 'harmonious balance with ... beautiful economy of means'. The continuing poverty of the Strangeways meant a great deal of making do and recycling: Fell commented to the Rockefeller Foundation in the early 1950s on the uses found for the packing of the equipment it had provided. Such meticulous attention to what might be dismissed as minutiae is perhaps stereotypically 'feminine'.

Fell was quietly feminist, being involved in organizations representing women in science and in education, addressing girls' schools on speech days on the delights of a scientific career, encouraging the career development of women in science and, although unmarried, sympathetic to the needs of women combining a career with marriage and motherhood.[33] An obvious factor in her attitude towards scientific work was relish. Her letters to Sir Edward Mellanby about their joint work on the effects of Vitamin A on skeletal development include such comments as 'On Sunday I shall have an orgy of staining slides'; 'this is a small cry of joy'; 'I had a lovely Saturday afternoon, with the whole lab to myself'; 'I am longing to get back to experiments again'.[34]

Such pleasure in work was not unique to Fell: 'working for pleasure, for the sheer joy of it' seems to have characterized crystallography well up until the 1960s.[35] Women may not have expected the worldly rewards available to men but they did have this almost luxurious sense of working at something which they found passionately absorbing and entirely fascinating. While women may have had few chances of attaining managerial and administrative posts, such posts would have involved sacrifice of work they were good at and loved. Gill Hudson's essay on Dorothy Hodgkin raises the additional issue of the desire for work to be of use which led Hodgkin to elucidate the crystalline structure of medically active substances such as Vitamin B12, insulin, and penicillin, mentioning the number of women in the 1930s who were working in the socially relevant field of nutrition.[36] The purposes for which women saw themselves as working in medicine or science need to be considered: how far was there a philanthropic agenda, at some level or another?

Recovering Stories

It is not necessarily easy to tease out the complex strands of these intricate tales. We must consider ourselves fortunate to find any biographical account more detailed than the dry bones of an obituary for many of these women, for example anything along the lines of Isabel Hutton's *Memories of a Doctor in War and Peace*; in some cases one would be glad of even the briefest obituary. Women's own papers often do not survive: few have had that confidence in their enduring historical significance which led Marie Stopes to bequeath three van-loads of papers to the British Museum.[37]

Records of a woman's career may lie concealed to a greater or lesser extent among those of a husband or colleague. May Mellanby's papers are intermingled with those of her husband, Sir Edward Mellanby.[38] Papers of Julia Bell (1879–1979), who began as a mathematician, did postgraduate work in astronomy, qualified in medicine and became a medical geneticist,[39] are among the papers of Karl Pearson, J.B.S. Haldane and L.S. Penrose, colleagues during her lengthy career at University College, London.[40] Collaboration may have been so close that papers are not readily separable, as with those of Elsie Widdowson, part of the team of 'McCance and Widdowson' responsible for 'the nutritionist's bible', *The Composition of Foods*,[41] or those of Innes Pearse, inextricably mingled with those of George Scott Williamson, co-founder of the Pioneer Health Centre at Peckham.[42] Sometimes a woman's papers are almost impossible to disentangle from those of the institution with which she was connected – papers of Letitia Fairfield lurk among the records of the Public Health Department of the London County Council in the Greater London Record Office, while detaching the papers of Honor Fell from those of the Strangeways Laboratory is rather like separating Siamese twins. In many cases a career can only be reconstructed through the records of the bodies or institutions with which a particular woman was involved, or the papers of her (usually male) associates.[43]

Sir Peter Medawar may have been sanguine in claiming in *Advice to a Young Scientist* that equality of treatment of the sexes was the norm, not the exception, in science: though possibly immunology approaches somewhat more nearly to that ideal than, say, nuclear physics. Nevertheless he made the important point that the young woman ambitious for a scientific career should look not to exceptional women like Marie Curie but to 'the tens of thousands of women gainfully and often happily engaged in scientific pursuits'.[44] This is precisely the point of recuperating this neglected area of women's history. To retrieve the careers of these forgotten women in science is as possible, worthwhile, and exciting as their work was for them.

Notes

1 V. Woolf (1928), *A Room of One's Own*, 1970 edition, Harmondsworth, Penguin, pp. 79–93.
2 J. Marcus (1987), *Virginia Woolf and the Languages of Patriarchy*, Bloomington, University of Indiana Press, p. 176; I am indebted to Angela Ingram for this reference.
3 M. Creese (1991), 'British Women of the Nineteenth and Early Twentieth Centuries who Contributed to Research in the Chemical Sciences', *British Journal for the History of Science*, 24, pp. 275–305; a highly illuminating study.
4 P. Toynbee, 'Appreciation: Dame Janet Vaughan', *The Guardian*, 13 Jan 1993.
5 M. Julian (1990), 'Women in Crystallography', in G. Kass-Simon and P. Farnes (Eds), *Women of Science: Righting the Record*, Bloomington, Indiana University Press, pp. 335–83; G. Hudson (1991), 'Unfathering the Thinkable: Gender, Science and Pacifism in the 1930s', in M. Benjamin (Ed.) *Science and Sensibility: Gender and Scientific Enquiry, 1780–1945*, Oxford, Basil Blackwell, pp. 264–86.

6   P. Abir-Am and D. Outram (Eds) (1987), *Uneasy Careers and Intimate Lives: Women in Science 1789–1979*, New Brunswick, Rutgers University Press; Benjamin (1991).

7   R. Dingley (1967), 'Hospitals Started by Women Doctors', *Journal of the Medical Women's Federation*, 49, pp. 108–11.

8   I. Hutton (1960), *Memories of a Doctor in War and Peace*, London, William Heinemann, p. 207.

9   *Ibid.*, p. 213.

10   *Ibid.*, pp. 218–20.

11   *Ibid.*, pp. 237–46; obituaries in *The Lancet*, 1960, i, pp. 231–2; *British Medical Journal*, 1960, i, pp. 353–4.

12   Obituaries in *British Medical Journal*, 1978, i, pp. 372–3; *The Lancet*, 1978, i, pp. 342–3. V. Glendinning (1987), *Rebecca West: A Life*, London, Weidenfeld and Nicolson, and R. West (1987), *Family Memories* (edited and introduced by F. Evans), London, Virago, shed oblique and unflattering light on Fairfield: the sisters' relationship was strained.

13   Archives of the Family Planning Association in the Contemporary Medical Archives Centre at the Wellcome Institute for the History of Medicine (CMAC): files on Dr Joan Malleson, CMAC: SA/FPA/A.14/58.1–2. It is perhaps amusing to note her comment on Isabel Hutton's *The Hygiene of Marriage* (1923): 'this old stager . . . is a shocker!', recommending (early 1950s) that it be dropped from the FPA recommended list.

14   The very large archive of the FPA at the CMAC would repay study from the point of view of careers of women doctors, both those, like Malleson, influential at the national level and those working at a local, clinic, level.

15   The archive of the Medical Women's Federation (CMAC) is disappointing as a record of what was going on for women in the medical profession at this period, though the journal of the Federation might provide useful leads. The British Medical Association gave some attention to the question of women in the medical profession ('Groups' files in CMAC).

16   P. Farnes (1990), 'Women in Medical Science', in Kass-Simon and Farnes, pp. 268–99; H. Viets (1965), 'The Miracle at St. Alfege's', *Medical History*, 9, pp. 184–5; *The Lancet*, 1974, ii, pp. 1401–2, 1582.

17   Poor Law and hospital records in the Greater London Record Office might shed further light on her career.

18   P. Phillips (1990), *The Scientific Lady: A Social History of Women's Scientific Interests, 1520–1918*, London, Weidenfeld and Nicolson.

19   Benjamin (1991), pp. 1–23.

20   Sir Edward and Lady Mellanby papers in the CMAC: PP/MEL/B.6; *Biographical Memoirs of Fellows of the Royal Society* (1955), 1, pp. 192–222.

21   Benjamin (1991), pp. 1–23.

22   Abir-Am and Outram (1987), pp. xi–xii.

23   Creese (1991).

24   *Ibid.*, citing Sayre's biography of Rosalind Franklin: Franklin may have been the victim of expectations of her colleagues not borne out in the chase for the 'Double Helix', as well as of gender attitudes of the 1950s.

25   Hopkins did not enter medical research through the standard routes but had previously been an analytical chemist; the Irish Bernal was also a Communist; C.J. Martin, who appointed so many women at the Lister Institute and made Harriette Chick a team-leader in nutritional biochemistry, went to work at 15 as a clerk in an insurance company and only studied medicine after matriculating at university through evening classes: Creese (1991); Hudson (1991); H. Chick *et al.* (1971), *War on Disease: A History of the Lister Institute*, London, Andre Deutsch, pp. 63–71.

26   'Jenner Memorial Scholarship: Appointment of Harriette Chick, 1905', Archives of the Lister Institute in the CMAC: SA/LIS/H.3.

27   Along with Dame Kathleen Lonsdale (1903–1971), the crystallographer.

28   Creese (1991).

29   *Beit Memorial Fellowships in Medical Research*, Dec 1962 (married status is only indicated when the name differs from that under which research was done); archives of the Beit Memorial Fellowships Trust in the CMAC.

30   E.M. Tansey (*c.* 1991), 'The Beardless Ladies of Physiology? An Historical Account of Women and the Physiological Society' in L. Bindman, A. Brading and T. Tansey (Eds) (1993), *Women Physiologists*, Colchester, Portland Press.

31   Chick *et al.* (1971).

32   Hudson (1991).

33   Most of this information was gleaned while sorting Fell's papers and those of the Strangeways Research Laboratory, and related collections in the CMAC; handlists are now available. L. Hall 'Illustrations from the Wellcome Institute Library: The Strangeways Research Laboratory' is forthcoming in *Medical History*. See also *Biographical Memoirs of Fellows of the Royal Society* (1987), 33, pp. 237–59, on Fell.

34   Sir Edward Mellanby's correspondence, 1954, CMAC: PP/MEL/B.39–40.

35   Julian (1990).

36   Hudson (1991).

37   L. Hall (1983), 'The Stopes Collection in the Contemporary Medical Archives Centre of the Wellcome Institute for the History of Medicine', *Bulletin of the Society for the Social History of Medicine*, 32, pp. 50–1.

38   Papers of Sir Edward and Lady (May) Mellanby in the CMAC.

39   G. Wolstenholme (Ed.) (1984), *Lives of the Fellows of the Royal College of Physicians (Munk's Roll): Vol VII: To 1983*, Oxford, IRL Press.

40   Information from Gill Furlong, Archivist, University College London.

41   Papers of Professor R.A. McCance and Elsie M. Widdowson in the CMAC.

42   Papers of George Scott Williamson and Innes Hope Pearse in the CMAC.

43   E.g. records of Sir Frederick Gowland Hopkins and J.D. Bernal in Cambridge University Library; those of the Braggs at the Royal Institution London.

44   P.B. Medawar (1979), *Advice to A Young Scientist*, London, Harper and Row, pp. 20–1.

Notes on Contributors

Johanna Alberti has taught history since 1962 in a variety of institutions including schools, polytechnics, and the Workers' Educational Association. She is currently employed by the Open University and the Universities of Newcastle and Northumbria. She is the author of *Beyond Suffrage: Feminists in War and Peace, 1914–1928* (Macmillan, 1989).

Hanna Behrend is a retired Senior Lecturer in English Literature and Women's Studies. She has published on historical and literary subjects, women writers, women in East Germany, etc., and was a founding member of the East German Independent Women's Association.

Sue Bruley has taught Historical Studies at the University of Portsmouth since 1988. She has numerous publications which relate to her interest in women in twentieth-century Britain. She has made extensive use of oral testimony in her work.

Katy Deepwell, artist/writer/lecturer, teaches art history at Goldsmiths' College, London University. She has curated *Ten Decades: The Careers of Ten Women Artists Born 1897–1906*, Norwich, April 1992. From 1991 to 1993 she was Chair of the Executive Committee of the Women Artists Slide Library, London.

Jacqueline de Vries was born in 1965. She received her MA and PhD degrees from the University of Illinois. She teaches Modern British and Comparative Women's History at Carleton College in Northfield, Minnesota, and is revising her thesis ' "A New Heaven and a New Earth": Feminism, Culture and Religion in Great Britain, 1900–1930', for publication.

Martin Durham was born in 1951 in Plymouth and is a Senior Lecturer in Politics at the University of Wolverhampton. He has written articles on the politics of the family in Britain and the USA and on women and British fascism, and he is the author of *Sex and Politics: The Family and Morality in the Thatcher Years* (Macmillan, 1991).

Elizabeth Edwards has been Librarian and Archivist at Homerton College, Cambridge, since 1984. She has published a number of articles on women's colleges and is particularly interested in women's experience in teacher training institutions.

Veronica Gillespie was born in 1914 and educated at Lady Margaret Hall, Oxford. She served in Searchlight regiments during the war, and worked as a freelance for IPC magazines.

Lesley A. Hall (born 1949) is Senior Assistant Archivist in the Contemporary Medical Archives Centre at the Wellcome Institute for the History of Medicine. Besides a Diploma in Archive Administration, she has a PhD in History of Medicine, and has published *Hidden Anxieties: Male Sexuality 1900–1950* (Polity, 1991).

Fred Hunter was born in 1934. A graduate of Cambridge University, he worked in the Civil Service until 1973, running the London Radio News service. He helped start commercial radio in 1973, running Independent Radio News; he moved to ITN in 1974, before entering academia. He was awarded The City University's journalism department's first PhD in 1984, for research into journalism education. He is now researching British women journalists 1840–1990.

Maroula Joannou was born in 1947. She has a PhD in English from Cambridge, and is Senior Lecturer in English Studies at Anglia University. She has contributed to *Rediscovering the Forgotten Radicals: Women Writers 1890–1940* (edited Daphne Patai and Angela Ingram) and is the author of *Towards a Women's Agenda in Writing: Feminist Consciousness and Women's Prose 1918–1938* (forthcoming). She is editing a volume of essays on Modern and Renaissance Studies in memory of Margot Heinemann with Andy Croft and David Margolies.

Brigitte Libmann was born in 1963 and studied Literature and Political Philosophy at the Sorbonne. She teaches Philosophy in Paris.

Elizabeth McClair was born in 1950. She graduated as a mature student from the University of Sussex in 1990. She is currently teaching in Further Education and writing.

Maggie Morgan was born in 1958. She is a Lecturer in History at the West Sussex Institute of Higher Education. After studying Cultural Studies at Portsmouth Polytechnic, as a mature student, she worked in adult education and for the Open University. Her contribution is based on research for her completed DPhil, held in the University of Sussex Library.

Sybil Oldfield was born in 1938 and educated at the Universities of New Zealand and London. She is now Senior Lecturer in English at the University of Sussex; her books include *Spinsters of this Parish* (Virago, 1984) and *Women Against the Iron Fist* (Basil Blackwell, 1989).

Gillian Scott was born in 1955, studied at Sussex University, completing a DPhil there on the Women's Co-operative Guild in 1988, and is currently Senior Lecturer in the School of Historical and Critical Studies at the (new) University of Brighton.

Christine Zmroczek was born in 1947. She has been involved in Women's Studies and feminist activity since *c.* 1973. She is a Lecturer in Women's Studies at Roehampton Institute and is Managing Editor of *Women's Studies International Forum*.

Appendix

Archive Resources for Research on 20th Century British Women

Fawcett Library, Old Castle Street, London E1 – the premier resource on all British women's issues.

Women's Art Library (formerly Women Artists Slide Library), Fulham Palace, Bishops Road, London SW6 6EA.

Imperial War Museum, Lambeth Road, London SE1 – for British women in WWI and WWII.

Friends' House Library, Euston Road, opposite Euston Station, London NW1 2BE – for Quaker women.

Contemporary Medical Archives Centre, Wellcome Institute for the History of Medicine, 183 Euston Road, London NW1 2BE – for material on medical women, and on maternal health, midwifery and nursing.

Wiener Library, Devonshire Street, London W1N 2BH – for Jewish refugees and refugee workers.

TUC Library, Congress House, Great Russell Street, London WC1 – for women in trade unions.

Labour History Library, Labour Party Head Quarters, Walworth Road, London – for labour party women, women's issues and labour policy.

County and City Public Record Offices hold papers on local prominent individuals and organisations as well as material on all local issues. Large city libraries, such as Birmingham, keep 'scrapbooks' of prominent citizens. (An exemplary listing on local resources for women's history is: *Resources for Women's History in Greater Manchester*, published by National Museum of Labour History, 103 Princess Street, Manchester M1 6DD.)

Universities and Medical Schools keep archives on distinguished alumni. Some universities, like Reading, hold special collections in women's history.

The British Library for Political and Economic Science, LSE, London, and the University of Hull hold papers of the Women's Co-operative Guild.

Lesbian Archive and Information Centre, BCM 7005, London WC1N 3XX.

The University Women's Club, Audley Square, London, holds papers on its former members. Monks House Papers, University of Sussex Library for Virginia Woolf papers.

Mass Observation Archive, University of Sussex Library for diaries and social observation reports by obscure 'ordinary' – and extraordinary – women since the late 1930s.

See also:

Margaret Barrow, *Women 1870–1928: A Select Guide to Printed and Archival Sources in the UK*, Mansell, 1981.

J. Foster & J. Sheppard *British Archives: A Guide to Archive Resources in the United Kingdom*, London Macmillan, 2nd ed, 1989.

E. Garrett, *Women in Economic & Social History, UK Research Directory*, CAMPOP, 27 Trumpington Street, Cambridge CB2 1QA.

Index